D1082986

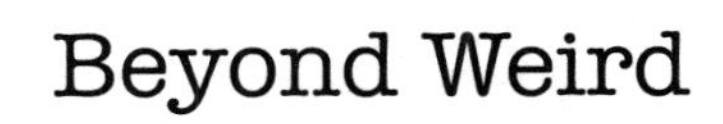
Beyond Weird

Beyond Weird

Why everything you thought you knew about quantum physics is different

PHILIP BALL

THE UNIVERSITY OF CHICAGO PRESS

The University of Chicago Press, Chicago 60637

Published 2018
Printed in the United States of America

27 26 25 24 23 22 21 20 19 18 1 2 3 4 5

ISBN-13: 978-0-226-55838-7 (cloth)
ISBN-13: 978-0-226-59498-9 (e-book)
DOI: https://doi.org/10.7208/chicago/9780226594989.001.0001

Originally published by The Bodley Head, 2018

LIBRARY OF CONGRESS CATALOGING-IN-PUBLICATION DATA
Names: Ball, Philip, 1962– author.
Title: Beyond weird : why everything you thought you knew about quantum physics is different / Philip Ball.
Description: Chicago : The University of Chicago Press, 2018. | Includes bibliographical references and index.
Identifiers: LCCN 2018008602 | ISBN 9780226558387 (cloth : alk. paper) | ISBN 9780226594989 (e-book)
Subjects: LCSH: Quantum theory—Popular works.
Classification: LCC QC174.123 .B36 2018 | DDC 530.12—dc23
LC record available at https://lccn.loc.gov/2018008602

♾ This paper meets the requirements of ANSI/NISO Z39.48-1992 (Permanence of Paper).

BEYOND WEIRD

By way of introduction . . .

To encounter the quantum is to feel like an explorer from a faraway land who has come for the first time upon an automobile. It is obviously meant for use, and important use, but what use?

John Archibald Wheeler

Somewhere in [quantum theory] the distinction between reality and our knowledge of reality has become lost, and the result has more the character of medieval necromancy than of science.

Edwin Jaynes

We must never forget that 'reality' too is a human word just like 'wave' or 'consciousness'. Our task is to learn to use these words correctly – that is, unambiguously and consistently.

Niels Bohr

[Quantum mechanics] is a peculiar mixture describing in part realities of Nature, in part incomplete human information about Nature – all scrambled up by Heisenberg and Bohr into an omelette that nobody has seen how to unscramble.

Edwin Jaynes

Arguably the most important lesson of quantum mechanics is that we need to critically revisit our most basic assumptions about nature.

Yakir Aharonov et al.

I hope you can accept Nature as she is – absurd.

Richard Feynman

No one can say what

quantum mechanics means
(and this is a book about it)

Richard Feynman said that in 1965. In the same year he was awarded the Nobel Prize in Physics, for his work on quantum mechanics.

In case we didn't get the point, Feynman drove it home in his artful Everyman style. 'I was born not understanding quantum mechanics,' he exclaimed merrily, '[and] I still don't understand quantum mechanics!' Here was the man who had just been anointed one of the foremost experts on the topic, declaring his ignorance of it.

What hope was there, then, for the rest of us?

Feynman's much-quoted words help to seal the reputation of quantum mechanics as one of the most obscure and difficult subjects in all of science. Quantum mechanics has become symbolic of 'impenetrable science', in the same way that the name of Albert Einstein (who played a key role in its inception) acts as shorthand for scientific genius.

Feynman clearly didn't mean that he couldn't *do* quantum theory. He meant that this was *all* he could do. He could work through the math just fine – he invented some of it, after all. That wasn't the problem. Sure, there's no point in pretending that the math is easy, and if you never got on with numbers then a career in quantum mechanics isn't for you. But neither, in that case, would be a career in fluid mechanics, population dynamics, or economics, which are equally inscrutable to the numerically challenged.

No, the equations aren't why quantum mechanics is perceived to be so hard. It's the ideas. We just can't get our heads around them. Neither could Richard Feynman.

His failure, Feynman admitted, was to understand what the math was saying. It provided numbers: predictions of quantities that could be tested against experiments, and which invariably survived those tests. But Feynman couldn't figure out what these numbers and equations were really about: what they said about the 'real world'.

One view is that they don't say anything about the 'real world'. They're just fantastically useful machinery, a kind of black box that we can use, very reliably, to do science and engineering. Another view is that the notion of a 'real world' beyond the math is meaningless, and we shouldn't waste our time thinking about it. Or perhaps we haven't yet found the right math to answer questions about the world it purports to describe. Or maybe, it's sometimes said, the math tells us that 'everything that can happen does happen' – whatever *that* means.

This is a book about what quantum math really means. Happily, we can explore that question without having to look very deeply into the math itself. Even what little I've included here can, if you prefer, be gingerly set aside.

I am not saying that this book is going to give you the answer. We don't have an answer. (Some people do have an answer, but only in the sense that some people have the Bible: their truth rests on faith, not proof.) We do, however, now have better questions than we did when Feynman admitted his ignorance, and that counts for a lot.

What we *can* say is that the narrative of quantum mechanics – at least among those who think most deeply about its meaning – has changed in remarkable ways since the end of the twentieth century. Quantum theory has revolutionized our concept of atoms, molecules, light and their interactions, but that transformation didn't happen abruptly and in some ways it is still happening now. It began in the early 1900s and it had a workable set of equations and ideas by the late 1920s. Only since the 1960s, however, have we begun to glimpse what is most fundamental and important about the theory, and some of the crucial experiments have been feasible only from the 1980s. Several of them have been performed in the twenty-first century. Even today we are still trying to get to grips with the central ideas, and are still testing their limits. If what we truly want is a theory that is well understood rather than simply one that does a good job at calculating numbers, then we still don't really have a quantum theory.

This book aims to give a sense of the current best guesses about what that *real* quantum theory might look like, if it existed. It rather seems as though such a theory would unsettle most if not all we take for granted about the deep fabric of the world, which appears to be a far stranger and more challenging place than we had previously envisaged. It is not a place where different physical rules apply, so much as a place where we are forced to rethink our ideas about what we mean by a physical world and what we think we are doing when we attempt to find out about it.

In surveying these new perspectives, I want to insist on two things that have emerged from the modern renaissance – the word is fully warranted – in investigations of the foundations of quantum mechanics.

First, what is all too frequently described as the *weirdness* of quantum physics is not a true oddity of the quantum world but comes from our (understandably) contorted attempts to find pictures for visualizing it or stories to tell about it. Quantum physics defies intuition, but we do it an injustice by calling that circumstance 'weird'.

Second – and worse – this 'weirdness' trope, so nonchalantly paraded in popular and even technical accounts of quantum theory, actively obscures rather than expresses what is truly revolutionary about it.

Quantum mechanics is in a certain sense not hard at all. It is baffling and surprising, and right now you could say that it remains cognitively impenetrable. But that doesn't mean it is *hard* in the way that car maintenance or learning Chinese is hard (I speak with bitter experience of both). Plenty of scientists find the theory easy enough to accept and master and use.

Rather than insisting on its difficulty, we might better regard it as a beguiling, maddening, even amusing gauntlet thrown down to challenge the imagination.

For *that* is indeed what is challenged. I suspect we are, in the wider cultural context, finally beginning to appreciate this. Artists, writers, poets and playwrights have started to imbibe and deploy ideas from quantum physics: see, for instance, plays such as Tom Stoppard's *Hapgood* and Michael Frayn's *Copenhagen*, and novels such as Jeanette Winterson's *Gut Symmetries* and Audrey Niffenegger's *The Time Traveler's Wife*. We can argue about how accurately or aptly these writers appropriate the scientific ideas, but it is right that there should be imaginative responses to quantum mechanics, because it is quite possible that only an imagination sufficiently broad and liberated will come close to articulating what it is about.

There's no doubt that the world described by quantum mechanics defies our intuitions. But 'weird' is not a particularly useful way to talk about it, since that world is also our world. We now have a fairly good, albeit still incomplete, account of how the world familiar to us, with objects having well-defined properties and positions that don't depend on how we choose to measure them, emerges from the quantum world. This 'classical' world is, in other words, a special case of quantum theory, not something distinct from it. If anything deserves to be called weird, it is us.

•

Here are the most common reasons for calling quantum mechanics weird. We're told it says that:

- Quantum objects can be both waves and particles. This is *wave-particle duality.*
- Quantum objects can be in more than one state at once: they can be both *here* and *there*, say. This is called *superposition*.
- You can't simultaneously know exactly two properties of a quantum object. This is *Heisenberg's uncertainty principle*.
- Quantum objects can affect one another instantly over huge distances: so-called 'spooky action at a distance'. This arises from the phenomenon called *entanglement*.
- You can't measure anything without disturbing it, so the human observer can't be excluded from the theory: it becomes unavoidably subjective.
- Everything that can possibly happen does happen. There are two separate reasons for this claim. One is rooted in the (uncontroversial) theory called quantum electrodynamics that Feynman and others formulated. The other comes from the (extremely controversial) 'Many Worlds Interpretation' of quantum mechanics.

Yet quantum mechanics says none of these things. In fact, quantum mechanics doesn't say *anything* about 'how things are'. It tells us what to expect when we conduct particular experiments. All of the claims above are nothing but *interpretations* laid on top of the theory. I will ask to what extent they are good interpretations (and try to give at least a flavour of what 'interpretation' might mean) –

but I will say right now that none of them is a very good interpretation and some are highly misleading.

The question is whether we can do any better. Regardless of the answer, we are surely being fed too narrow and too stale a diet. The conventional catalogue of images, metaphors and 'explanations' is not only clichéd but risks masking how profoundly quantum mechanics confounds our expectations.

It's understandable that this is so. We can hardly talk about quantum theory at all unless we find stories to tell about it: metaphors that offer the mind purchase on such slippery ground. But too often these stories and metaphors are then mistaken for the way things are. The reason we can express them at all is that they are couched in terms of the quotidian: the quantum rules are shoehorned into the familiar concepts of our everyday world. But that is precisely where they no longer seem to fit.

•

It's very peculiar that a scientific theory should demand interpretation at all. Usually in science, theory and interpretation go together in a relatively transparent way. Certainly a theory might have *implications* that are not obvious and need spelling out, but the basic *meaning* is apparent at once.

Take Charles Darwin's theory of evolution by natural selection. The objects to which it refers – organisms and species – are relatively unambiguous (if actually a little challenging to make precise), and it's clear what the theory says about how they evolve. This evolution depends on two ingredients: random, inheritable mutations in traits; and competition for limited resources that gives a reproductive advantage to individuals with certain variants of a trait.

How this idea plays out in practice – how it translates to the genetic level, how it is affected by different population sizes or different mutation rates, and so on – is really rather complex, and even now not all of it is fully worked out. But we don't struggle to understand what the theory *means*. We can write down the ingredients and implications of the theory in everyday words, and there is nothing more that needs to be said.

Feynman seemed to feel that it was impossible and even pointless to attempt anything comparable for quantum mechanics:

> We can't pretend to understand it since it affronts all our commonsense notions. The best we can do is to describe what happens in mathematics, in equations, and that's very difficult. What is even harder is trying to decide what the equations mean. That's the hardest thing of all.

Most users don't worry too much about these puzzles. In the words of the physicist David Mermin of Cornell University, they 'shut up and calculate'.* For many decades quantum theory was regarded primarily as a mathematical description of phenomenal accuracy and reliability, capable of explaining the shapes and behaviours of molecules, the workings of electronic transistors, the colours of nature and the laws of optics, and a whole lot else. It would be routinely described as 'the theory of the atomic world':

* It's commonly but wrongly believed that Feynman said this. The belief was so widespread that at one point even Mermin began to fear his quip might in fact have been unconsciously echoing Feynman. But Feynman was not the only physicist with a smart line in quantum aphorisms, as we shall see.

an account of what the world is like at the tiniest scales we can access with microscopes.

Talking about the interpretation of quantum mechanics was, on the other hand, a parlour game suitable only for grandees in the twilight of their career, or idle discussion over a beer. Or worse: only a few decades ago, professing a serious interest in the topic could be tantamount to career suicide for a young physicist. Only a handful of scientists and philosophers, idiosyncratically if not plain crankily, insisted on caring about the answer. Many researchers would shrug or roll their eyes when the 'meaning' of quantum mechanics came up; some still do. 'Ah, nobody understands it anyway!'

How different this is from the attitude of Albert Einstein, Niels Bohr and their contemporaries, for whom grappling with the apparent oddness of the theory became almost an obsession. For them, the meaning mattered intensely. In 1998 the American physicist John Wheeler, a pioneer of modern quantum theory, lamented the loss of the '*desperate* puzzlement' that was in the air in the 1930s. 'I want to recapture that feeling for all, even if it is my last act on Earth', Wheeler said.

Wheeler may indeed have had some considerable influence in making this deviant tendency become permissible again, even fashionable. The discussion of options and interpretations and meanings may no longer have to remain a matter of personal preference or abstract philosophizing, and if we can't say what quantum mechanics means, we can now at least say more clearly and precisely what it does *not* mean.

This re-engagement with 'quantum meaning' comes partly because we can now do experiments to probe foundational issues that were previously expressed as mere thought experiments and considered to be on the border of metaphysics: a mode of thinking that, for better or worse, many scientists disdain. We can now put quantum paradoxes and puzzles to the test – including the most famous of them all, Schrödinger's cat.

These experiments are among the most ingenious ever devised. Often they can be done on a benchtop with relatively inexpensive equipment – lasers, lenses, mirrors – yet they are extraordinary feats to equal anything in the realm of Big Science. They involve capturing and manipulating atoms, electrons or packets of light, perhaps one at a time, and subjecting them to the most precise examination. Some experiments are done in outer space to avoid the complications introduced by gravity. Some are done at temperatures colder than the void between the stars. They might create completely new states of matter. They enable a kind of 'teleportation'; they challenge Werner Heisenberg's view of uncertainty; they suggest that causation can flow both forwards and backwards in time or be scrambled entirely. They are beginning to peel back the veil and show us what, if anything, lies beneath the blandly reassuring yet mercurial equations of quantum mechanics.

Such work is already winning Nobel Prizes, and will win more. What it tells us above all else is very clear: the apparent oddness, the paradoxes and puzzles of quantum mechanics, are real. We cannot hope to understand how the world is made up unless we grapple with them.

Perhaps most excitingly of all, because we can now do experiments that exploit quantum effects to make possible what sounds as though it should be impossible, we can put those tricks to work. We are inventing quantum technologies that can manipulate information in unprecedented ways, transmit secure information that cannot be read surreptitiously by eavesdroppers, or perform calculations that are far beyond the reach of ordinary computers. In this way more than any other, we will all soon have to confront the fact that quantum mechanics is not some weirdness buried in remote, invisible aspects of the world, but is our current best shot at uncovering the laws of nature, with consequences that happen right in front of us.

What has emerged most strongly from this work on the fundamental aspects of quantum theory over the past decade or two is that it is not a theory about particles and waves, discreteness or uncertainty or fuzziness. It is a theory about *information*. This new perspective gives the theory a far more profound prospect than do pictures of 'things behaving weirdly'. Quantum mechanics seems to be about what we can reasonably call a view of reality. More even than a question of 'what can and can't be known', it asks what a *theory of knowability* can look like.

I've no intention of hiding it from you that this picture doesn't resolve the ways quantum mechanics challenges our intuition. It seems likely that nothing can do that. And talking about 'quantum information' brings its own problems, because it raises questions about what this information *is* – or what it is about, because information is not a thing that you can point to in the way you can with an apple or even (in some cases) with an atom. When we use the word 'information' in everyday usage it is bound up with considerations of language and meaning, and thus of context. Physicists have a definition of information that doesn't match this usage – it is greatest when most random, for example – and there are difficult issues about how, in quantum mechanics, such a recondite definition impinges on the critical issue of *what we know*. So we don't have all the answers. But we do have better questions, and that's some kind of progress.

•

You can see that I'm already struggling to find a language that works for talking about these things. That's OK, and you'll have to get used to it. That's how it should be. When words come too easily, it's because we haven't delved deeply enough (you'll see that scientists can be guilty of that too). 'We are suspended in language', said Bohr, who thought more profoundly about quantum mechanics than any of his contemporaries, 'in such a way that we cannot say what is up and what is down.'

It's almost an in-joke that popular accounts of quantum mechanics abound with statements along the lines of 'This isn't a perfect analogy, but ...' Then what typically follows is a visualization involving marbles and balloons and brick walls and the like. It is the easiest thing in the world for the pedant to say 'Oh, it's not really like that at all.' This isn't my intention. Such elaborately prosaic imagery is often a good place to start the journey, and I will sometimes resort to it myself. Sometimes an imperfect analogy like this is all that can be sensibly expected without engaging in detailed mathematical expositions, and even specialists sometimes have to entertain such pictures if they aren't ready to capitulate to pure abstraction. Richard Feynman did so, and that is good enough for me.

It's only when we abandon those mental crutches, however, that we can start to see why we need to take quantum mechanics more seriously. I don't mean that we should all be terribly earnest about it (Feynman wasn't), but that we should be prepared to be *much more unsettled* about it. I have barely scratched the surface, and *I* am unsettled. Bohr, again, understood this point. He once gave a talk on quantum mechanics to a group of philosophers, and was disappointed and frustrated that they sat and meekly accepted what he said rather than protesting vehemently. 'If a man does not feel dizzy when he first learns about the quantum of action [that is, quantum theory],' said Bohr, 'he has not understood a word.'

I'm suggesting that we don't worry enough about what quantum theory means. I don't mean that we're not interested – it's a peculiar fact that articles about the quirks of quantum theory in popular-science magazines and forums are almost invariably among the most widely read, and there are plenty of accessible books on the subject.* So why complain that we don't worry enough?

Because the issue is often made to seem like 'not *our* problem'. Reading about quantum theory often feels a little like reading anthropology: it tells of a far-off land where the customs are strange. We're comfortable enough about how *our* world behaves; it's this other one that's 'weird'.

That, however, is as parochial, if not quite as offensive, as if I were to assert that the customs of a tribe of New Guinea were 'weird' because they are not mine. Besides, it underestimates quantum mechanics. For one thing, the

* Many are excellent, but you could hardly do better than start with Jim Al-Khalili's new 'Ladybird Expert' book *Quantum Mechanics*.

more we understand about it, the more we appreciate how our familiar world is not distinct from it but a consequence of it. What's more, if there is a more 'fundamental' theory underlying quantum mechanics, it seems that it will have to retain the essential features that make the quantum world look so strange to us, extending them into new regimes of time and space. It is probably quantum all the way down.

Quantum physics implies that the world comes from a quite different place than the conventional notion of particles becoming atoms becoming stars and planets. All that happens, surely: but the fundamental fabric from which it sprang is governed by rules that defy traditional narratives. It is another quantum cliché to imply that those rules undermine our ideas of 'what is real' – but this, at least, is a cliché that we might usefully revisit with fresh eyes. The physicist Leonard Susskind is not exaggerating when he says that 'in accepting quantum mechanics, we are buying into a view of reality that is radically different from the classical view'.

Note that: a different view of *reality*, not a different kind of physics. If different physics is 'all' you want, you can look (say) to Einstein's theories of special and general relativity, in which motion and gravity slow time and bend space. That's not easy to imagine, but I reckon you can do it. You just need to imagine time passing more slowly, distances contracting: distortions of your grid references. You can put those ideas into words. In quantum theory, words are blunt tools. We give names to things and processes, but those are just labels for concepts that cannot be properly, accurately expressed in any terms but their own.

A different view of reality, then: if we're serious about that, we're going to need some philosophy. Many scientists, like many of us, take a seemingly pragmatic but rather naïve view of 'reality': it's just the stuff out there that we can see and touch and influence. But philosophers – from Plato and Aristotle through to Hume, Kant, Heidegger and Wittgenstein – have long recognized that this is to take an awful lot for granted that we really should interrogate more closely. Attempts to interpret quantum mechanics demand that interrogation, and so force science to take seriously some questions that philosophers have debated with great depth and subtlety for millennia: What is real? What is knowledge? What is existence? Scientists often have a tendency to respond to such questions with Johnsonian impatience, as though they are either self-evident or useless sophistry. But evidently they are not, and some quantum physicists are now happy to consider what philosophers had and have to say about them. And the field of 'quantum foundations' is the better for it.

•

Are we doomed, though, to be forever 'suspended in language', as Bohr said, not knowing up from down? Some researchers optimistically think that, on the contrary, it might eventually be possible to express quantum theory in terms of – as one of them has put it – 'a set of simple and physically intuitive principles, and a convincing story to go with them'. Wheeler once claimed that if we really understood the central point of quantum theory, we ought to be able to state it in one simple sentence.

Yet there is no guarantee, nor indeed much likelihood, that future experiments are going to strip away all the counter-intuitive aspects of quantum theory and reveal something as concrete, 'commonsensical' and satisfying as the old-style classical physics. Indeed, it is possible that we might *never* be able to say what quantum theory 'means'.

I have worded that sentence carefully. It's not exactly (or necessarily) that no one will *know* what the theory means. Rather, we might find our words and concepts, our ingrained patterns of cognition, to be unsuited to articulating a meaning worthy of the name. David Mermin expressed this adeptly when describing how many quantum physicists feel about Niels Bohr himself, who acquired the reputation of a guru with a quasi-mystical understanding that leaves physicists even now poring over his maddeningly cryptic words. 'I have been getting sporadic flashes of feeling that I may actually be starting to understand what Bohr was talking about', wrote Mermin:

> Sometimes the sensation persists for many minutes. It's a little like a religious experience and what really worries me is that if I am on the right track, then one of these days, perhaps quite soon, the whole business will suddenly become obvious to me, and from then on I will know that Bohr was right but be unable to explain why to anybody else.

It might be, then, that all we can ever do is shut up and calculate, and dismiss the rest as a matter of taste. But I think we can do better, and that we should at least aspire to. Perhaps quantum mechanics pushes us to the limits of what we can know and comprehend. Well then, let's see if we can push back a little.

Quantum mechanics is not

really about the quantum

The temptation to tell quantum mechanics as a historical saga is overwhelming. It's such a great tale. How, at the beginning of the twentieth century, physicists began to realize that the world is constructed quite differently from how they had supposed. How this 'new physics' began to disclose increasingly odd implications. How the founders puzzled, argued, improvised, guessed, in their efforts to come up with a theory to explain it all. How knowledge once deemed precise and objective now seemed uncertain, contingent and observer-dependent.

And the cast! Albert Einstein, Niels Bohr, Werner Heisenberg, Erwin Schrödinger, and other colourful intellectual giants like John von Neumann, Richard Feynman and John Wheeler. Best of all for its narrative value is the largely good-natured but trenchant dispute that rumbled on for decades between Einstein and Bohr about what it all meant – about the nature of reality. This is indeed a superb story, and if you haven't heard it before then you should.*

Yet most popular descriptions of quantum theory have been too wedded to its historical evolution. There is no reason to believe that the most important aspects of the theory are those that were discovered first, and plenty of reason to think that they are not. Even the term 'quantum' is something of a red herring, since the fact that the theory renders a description of the world granular and

* I'd recommend beginning with Manjit Kumar's *Quantum* (2008).

particulate (that is, divided into discrete *quanta*) rather than continuous and fluid is more a symptom than a cause of its underlying nature. If we were naming it today, we'd call it something else.

I'm not going to ignore this history. One simply cannot do that when discussing quantum mechanics, not least because what some of the historical doyens had to say on the matter – Bohr and Einstein in particular – remains perceptive and relevant today. But telling quantum theory chronologically can become a part of the problem that we have with it. It yokes us to a particular view of what matters – a view that no longer seems to be looking from the right direction.

•

Quantum theory had the strangest genesis. Its pioneers made it up as they went along. What else could they do? It was a new kind of physics – they couldn't deduce it from the old variety, although they were able nonetheless to commandeer a surprising amount of traditional physics and math. They cobbled old concepts and methods together into new forms that were often nothing much more than a wild guess at what kind of equation or mathematics might do the job.

It is extraordinary how these hunches and suppositions about very specific, even recondite, phenomena in physics cohered into a theory of such scope, precision and power. Far too little is made of this when the subject is taught, either as science or as history. The student is (certainly *this* student was) presented with the mathematical machinery as though it were a result of rigorous deduction and decisive experiment. No one tells you that it often lacks any justification beyond the mere (and obviously important) fact that it works.

Of course, this can't have been sheer luck. The reason Einstein, Bohr, Schrödinger, Heisenberg, as well as Max Born, Paul Dirac, Wolfgang Pauli and others, were able to concoct a mathematical quantum mechanics is that they possessed extraordinary physical intuition informed by their erudition in classical physics. They had amazing instincts for which pieces of conventional physics to use and which to throw away. This doesn't alter the fact that the formalism of quantum theory is makeshift and in the end rather arbitrary. Yes, the most accurate physical theory we possess is something of a Heath Robinson (Americans will say a Rube Goldberg) contraption. Worse than that – for those devices have a clear logic to their operation, a rational connection between one part and another. But most of the fundamental equations and concepts of quantum mechanics are (inspired) guesses.

•

Scientific discovery often starts with an observation or an experiment that no one can explain, and quantum mechanics was like that too. Indeed, the theory could surely not have arisen except from experiment – for there is absolutely no logical reason to expect anything it says. We can't reason ourselves into quantum theory (which, if we believe Jonathan Swift's famous aperçu, presumably means we can never reason ourselves out of it). It is simply an attempt to describe what we see when we examine nature closely enough.

What distinguishes quantum mechanics from other empirically motivated theories, however, is that the quest for underlying causes doesn't allow – at least, hasn't yet allowed – the theory to be constructed from more fundamental ingredients. With any theory, at some point you can't help asking 'So *why* are things this way? Where do these rules come from?' Usually in science you can answer those questions by careful observation and measurement. With quantum mechanics it is not so simple. For it is not so much a theory that one can test by observation and measurement, but a theory about what it means to observe and measure.

Quantum mechanics started as a makeshift gambit by the German physicist Max Planck in 1900. He was studying how objects radiate heat, which seemed like as conventional and prosaic a question as you could imagine a physicist asking. It was, to be sure, a matter of great interest to late-nineteenth-century physicists, but it scarcely seemed likely to require an entire new world view.

Warm objects emit radiation. If they are hot enough, some of that radiation is visible light: they become 'red hot', or with more heating, 'white hot'. Physicists devised an idealized description of this situation in which the emitting object is called a black body – which might sound perverse, but it just means that the object absorbs perfectly all radiation that falls on it. That keeps the issue simple: you only need focus on what gets emitted.

It was possible to make objects that behaved like black bodies – a hole in a warm oven would do the job – and measure how much energy they radiated at different wavelengths of light.* But explaining these measurements in terms of the vibrations within the warm body – the source of the emitted radiation – was no easy matter.

That explanation depended on how heat energy was distributed among the various vibrations. It was a problem for the science called thermodynamics, which describes how heat and energy are moved around. We now identify the vibrations of the black body with the oscillations of its constituent atoms. But when Planck studied the problem in the late nineteenth century there was still no direct evidence for the very existence of atoms, and he was vague about what the 'oscillators' are.

What Planck did seemed so very innocuous. He found that the discrepancy between what thermodynamic theory predicted for black-body radiation and what was seen experimentally could be eased by assuming that the energy of an oscillator can't just have *any* value, but is restricted to chunks of a particular size ('quanta') proportional to the frequency of oscillation. In other words, if an oscillator has a frequency f, then its energy can only take values that are whole-number multiples of f, multiplied by some constant denoted h and now called Planck's constant. The energy can be equal to hf, $2hf$, $3hf$ and so on, but cannot take values in between. This implies that each oscillator can only emit (and absorb) radiation in discrete packets with frequency f, as it moves between successive energy states.

* According to classical physics, light is a wave of conjoined electrical and magnetic fields moving through space. The wavelength is the distance between successive peaks. Most light, like sunlight, consists of waves of many different wavelengths, although laser light typically has a very narrow band of wavelengths. This wave view of light was one of the first casualties of quantum theory, as we'll see.

This story is often told as an attempt by Planck to avoid the 'ultraviolet catastrophe': the prediction, according to classical physics, that warm bodies should emit ever more radiation as the wavelength gets shorter (that is, closer to the ultraviolet end of visible light's spectrum). That prediction, which implies – impossibly – that a warm object radiates an infinite amount of energy, follows from the assumption that the heat energy of the object is shared out equally between all its vibrations.

It's true that Planck's quantum hypothesis, by supposing that the vibrations can't just take *any* frequencies, avoids this inconvenient prediction. But that was never his motivation for it. He thought that his new formula for black-body radiation applied only at low frequencies anyway, whereas the ultraviolet catastrophe loomed specifically at high frequencies. The myth probably reflects a sense that quantum theory needed some urgent-sounding crisis to precipitate it. But it didn't, and Planck's proposal excited no controversy or disquiet until Albert Einstein insisted on making the quantum hypothesis a general aspect of microscopic reality.

In 1905 Einstein proposed that quantization was a real effect, not just some sleight of hand to make the equations work. Atomic vibrations really do have this restriction. Moreover, he said, it applies also to the energy of light waves themselves: their energy is parcelled up into packets, called photons. The energy of each packet is equal to h times the light's frequency (how many wave oscillations it makes each second).

Many of Einstein's colleagues, including Planck, felt that he was taking far too literally what Planck had intended only as a mathematical convenience. But experiments on light and its interactions with matter soon proved Einstein right.

So it was that quantum mechanics seemed at the outset to be about this notion of 'quantized energy': how it increases in steps, not smoothly, for atoms and molecules and light radiation. This, we're told, was the fundamental physical content of the early theory; the rest was added as a theoretical apparatus for handling it. That, however, is a little like saying that Isaac Newton's theory of gravitation was a theory of how comets move through the solar system. It was indeed the appearance of a comet in 1680 that prompted Newton to think about the shape of their paths and to formulate a law of gravity that explained them. But his gravitational theory is not *about* comets. It expresses an underlying principle of nature, of which cometary motion is one manifestation. Likewise, quantum mechanics is not really 'about' quanta: the chunking of energy is a fairly incidental (though initially unexpected and surprising) outcome of it. Quantization was just what alerted Einstein and his colleagues that something was up with classical physics. It was the telltale clue, and no more. We shouldn't confuse the clue with the answer.

Although both Planck and Einstein were rightly rewarded with Nobel Prizes for introducing the 'quantum', that step was simply a historical contingency that set the ball rolling.* Several other experiments in the

* The citation for Einstein's 1921 prize was cautiously worded, acknowledging how his work had helped to understand a phenomenon called the photoelectric effect, which drew on the notion of light quanta. At that point, the full implications for quantum theory were still considered too speculative to be granted such credit. Einstein was actually awarded the prize in 1922, since the 1921 prize was deferred for a year in the absence of nominations that were deemed worthy.

1920s and 30s could equally have kick-started quantum theory, had it not already been launched.

Put it this way: grant the rules of quantum mechanics and you must get quantization, but the reverse is not true. Quantization of energy could, in itself, conceivably be a phenomenon of classical physics. Suppose that nature just happens to be constructed in such a way that, at the smallest scales, energies have to be quantized: restricted to discrete values in a staircase of possibilities. That's surprising – we don't seem to have any reason to expect it (although it turns out to explain a lot of our direct experience, such as why grass is green) – but hey, why not? This could have been the end of the matter: nature is grainy at small scales. Einstein would have been happy with that.

The best illustration I know of that quantization is rather incidental to quantum theory is found in the book *Quantum Mechanics: The Theoretical Minimum*, based on a series of lectures that Leonard Susskind, a professor of theoretical physics at Stanford University, gave to undergraduates, which were written up with the help of the writer Art Friedman. The book is described as being 'for anyone who ever regretted not taking physics at university, who knows a little but would like to know more'. That's a rather optimistic assessment, but with a reasonable level of math you could learn all you needed to know in this marvellous tract about the theory. With that aim in mind, Susskind has organized the material so as to tell you what you need to know in the order that you need to know it, in distinction from the common practice of introducing topics and concepts in more or less chronological order. When, then, do you learn about quantization of Planck's 'oscillators'? In the last chapter. In fact 'The Importance of Quantization' is the last section of that final chapter. That's how modern physics judges the *conceptual* significance of Planck's hypothesis, and it's a fair assessment.

•

So if you want to understand what quantum mechanics is really about, what *do* you start with? Susskind's Lecture 1 is 'Systems and Experiments'. Here Susskind explains what is fundamentally different between quantum and classical mechanics. And it's *not* (as is often implied too) that quantum works at small scales and classical at big ones.

Practically speaking, that often *is* the difference, but only because, as we will see later, by the time objects get as big as tennis balls, quantum rules have conspired to generate classical behaviour. The significance of the size difference is not in terms of what objects do, but in terms of our perceptions. Because we haven't evolved to perceive quantum behaviour except in its limiting form of classical behaviour, we've had no grounds to develop an intuition for it. At least, that is probably part of the story; there may be more to it, as we'll also see.

The key distinctions between classical mechanics and quantum mechanics, in Susskind's view, are these:

- Quantum physics has 'different abstractions' – how objects are represented mathematically, and how those representations are logically related.
- Quantum physics has a different relationship between the state of a system and the result of a measurement on that system.

Don't worry about the first of these yet; regard it as analogous to saying that the concepts we use in physics are different from those we use in, say, literary theory or macroeconomics. It's no big deal.

You should worry about the second, though. In a sense, all of quantum theory's counter-intuitive nature (I am trying very hard not to call it weirdness) is packaged up here.

What does it mean to talk about the relationship between the state of a system and a measurement on the system? It's an odd phrase, and that's because this relationship is usually so trivial that we don't even think about it. If a tennis ball is in the state of travelling through the air at 100 mph, and I measure its speed, then that is the value I measure. A measurement tells us about the state of the ball's motion. Of course, there are limits of accuracy – I might have to say that the speed is 100 ± 1 mph – but that's just an instrumental issue. I could probably do better.

So we have no problem saying that the tennis ball was travelling at 100 mph and then I measured it. The tennis ball had the pre-existing property of a speed of 100 mph, which I could determine by measurement. We would never think of saying that it was travelling at 100 mph *because* I measured it. That wouldn't make any sense.

In quantum theory, we do have to make statements like that. And then we can't help asking what it means. That's when the arguments start.

Later we'll see some of the concepts that have been developed to talk about this problem of measurement – of the relationship between the state of a quantum system and what we observe of it. We'll hear about the talismanic conceptual paraphernalia of quantum theory: wavefunctions, superposition, entanglement and so forth. But these are all just handy tools that enable us to make predictions about what a measurement will show us, which is by and large the goal of fundamental science.

That Susskind's second principle – the relationships between states and measurements – can be put into words, without any need for equations or fancy jargon, should reassure us. It's not easy to understand what the words mean, but they reflect the fact that the most fundamental message of quantum theory isn't a purely mathematical one.

Some physicists might be tempted to argue precisely the opposite: that the math *is* the most fundamental description. They might say this basically because the math makes perfect sense whereas the words don't quite. But that would be to make a semantic error: equations purportedly about physical reality are, without interpretation, just marks on paper. We can't hide behind equations from that 'not quite' – not if we truly want to derive *meaning*. Feynman knew this.

Susskind's second principle is really a statement about our active involvement with the world as we seek knowledge about it. *That* – which has been the bedrock of philosophy for over two millennia – is where we must look for meaning.

Quantum objects are
particle (but sometimes

neither wave nor

they might as well be)

One of the problems in talking about quantum objects is deciding what to call them. It seems like a trivial point, but actually it's fundamental.

'Quantum objects' is terribly clunky, and vague too. What's wrong with 'particle'? When we speak of electrons and photons, atoms and molecules, it seems perfectly reasonable to use that word, and I'll occasionally do so. Then we might have the image of a tiny little *thing*, a microscopic ball-bearing all hard and shiny. But probably the most widely known fact of quantum mechanics is that 'particles can be waves'. What then becomes of our compact little balls?

We could simply give these quantum things a new name: quantons, say, which by definition can show wave-like or particle-like behaviour. But there is more than enough jargon in this subject already, and replacing familiar, comfortable words with neologisms that seem designed only to sweep complications under the carpet doesn't feel terribly satisfactory. So for the present purposes, 'objects' and 'particles' will have to do. Except, I suppose, when they're like waves.

The notion of wave–particle duality goes back to the earliest days of quantum mechanics, but it is as much an impediment as it is a crutch to our understanding. Einstein expressed it by saying that quantum objects present us with a choice of languages, but it's too easily forgotten that this is *precisely* what it is: a struggle to formulate the right words, not a description of the reality behind them. Quantum objects are not sometimes particles and sometimes waves, like a football fan changing her team allegiance according to last week's results. Quantum objects are what they are, and we have no reason to suppose that 'what they are' changes in any meaningful way depending on how we try to look at them. Rather, all we can say is that what we measure sometimes looks like what we would expect to see if we were measuring discrete little ball-like entities, while in other experiments it looks like the behaviour expected of waves of the same kind as those of sound travelling in air, or that wrinkle and swell on the sea surface. So the phrase 'wave–particle duality' doesn't really refer to quantum objects at all, but to the interpretation of experiments – which is to say, to our human-scale view of things.

•

In 1924 the French physicist and aristocrat Louis de Broglie proposed that quantum particles – then still envisaged as tiny lumps of *stuff* – might display wave-like properties. His idea was, like so many others in early quantum theory, nothing much more than a hunch. He was generalizing from, indeed inverting, Einstein's earlier argument that light waves display particle-like behaviour when they manifest as photons with discrete energies.

If light waves can be particle-like, de Broglie said in his doctoral thesis, then might not the entities we've previously considered as particles (such as electrons) be wavy? The proposal was controversial and all but dismissed until Einstein suggested, after some reflection, that it was worth heeding after all. 'It looks completely crazy,' he wrote, 'but it's a completely sound idea.'

De Broglie didn't develop his idea into a full-blown theory. But there was already a mature mathematical theory of classical waves; maybe we could use that to describe the alleged waviness of particles? That's just what Erwin Schrödinger, a professor of physics at Zurich, did. After being given de Broglie's thesis and challenged to describe wave-like particles in formal terms, he wrote down an expression for how they might behave.

It was not quite like an ordinary wave equation of the sort used to describe water waves or sound waves. But it was mathematically very similar.

Why *wasn't* it identical? Schrödinger didn't explain his reason, and it now seems clear that he didn't exactly have one. He simply wrote down what he thought a wave equation for a particle such as an electron ought to look like. That he seems to have made such a good guess is even now rather extraordinary and mysterious. Or to put it another way: Schrödinger's wave equation, which is now a part of the core conceptual machinery of quantum mechanics, was built partly by intuition and imagination, albeit combined with a deeply informed sense of which parts of classical physics it was appropriate to commandeer. It can't be proved, but only inferred by analogy and good instinct. This doesn't mean that the equation is wrong or untrustworthy; but its genesis shows how creativity in science depends on more than cold reason.

Wave equations stipulate what the *amplitude* of the wave is in different parts of space. For a water wave, the amplitude is simply how high the water surface is. For a sound wave, it means how strongly the air is compressed in the peaks of the wave, and how severely it is 'stretched' or rarefied in the troughs. Pick a spot in space and you'll see the amplitude there change over time – big, then small, then big again – as the wave undulates across it.

But what is the 'amplitude' of an electron wave? Schrödinger guessed that it corresponds to the amount of electrical charge at that point in space, since each electron bears a single unit – a quantum – of electrical charge.

It was a natural thing to assume, but it was wrong. The wave in Schrödinger's equation isn't a wave of electron charge density. In fact it's not a wave that corresponds to any concrete physical property. It is just a mathematical abstraction – for which reason it is not really a wave at all, but is called a *wavefunction*.

It does, however, have a meaning. The German physicist Max Born argued that the amplitude of the wavefunction squared (amplitude × amplitude) indicates a *probability*. Specifically, from the value of the wavefunction at some position x, you can use Born's rule to calculate the chance that, if you perform an experiment to measure where the particle is, you will find it at x. Crudely speaking, if the amplitude of an electron wavefunction at x is 1 (in some units), and at y it is 2, then repeated experiments to determine the electron's position will find it at y four times (2×2) more often than at x.

How did Born know this? He didn't. Again, he 'guessed' (and again, drawing on a wealth of physical intuition). And as with the Schrödinger equation itself, we still have no fundamental way of *deriving* Born's rule. (Some researchers claim to have done so, but no such derivation is universally accepted.)

The Schrödinger equation, then, is an expression for finding out how an abstract entity called a wavefunction is distributed in space and how it evolves in time. And – here's the really important thing – this wavefunction contains *all the information one can possibly access* about the corresponding quantum particle. Once you have the particle's wavefunction, you can extract that information by doing something to it. For example, you can square it to find out the probability of finding the particle at any location in space.

The French physicist Roland Omnès put it nicely when he called the wavefunction 'the fuel of a machine that manufactures probabilities'. In general, the chance of measuring any particular value of an observable property of a quantum system in an experiment can be calculated by a particular mathematical manipulation of its wavefunction. The wavefunction encodes this information, and quantum math lets you extract it. There's a particular operation you conduct on the wavefunction to find a particle's momentum (mass × velocity), another operation to find its energy, and so on. In each case, what you get from this operation is *not* exactly the momentum, or energy, or whatever, that you'd measure in an experiment; it's the *average* value you'd expect to get from many such measurements.

Solving the Schrödinger equation to deduce a wavefunction is impossible to do exactly with pen and paper for anything but the simplest and most idealized systems. But there are ways of getting an *approximate* wavefunction for more complicated systems, like a molecule with many atoms. Once you have a good enough wavefunction, you can then use it to calculate all manner of properties: how the molecule will vibrate, how it will absorb light, how it will interact with other molecules.

Quantum mechanics gives you a mathematical prescription for doing those calculations, and when you've learnt how to handle the quantum calculus then you're off and running. The math of quantum mechanics looks rather fearsome, involving as it does imaginary numbers and calculus and things called projection operators. But it's really just a set of rules that describe how quantum states deliver expectations about specific results when we make measurements on them: a mechanism for reaching into a wavefunction and pulling out quantities that have the potential to be observed in experiments.

And you don't, unless you are so inclined, ever need to think about what it all 'means'. You can shut up and calculate.

•

There's no harm in that. But relying on wavefunctions for everything we can say or find out about quantum objects has some consequences that look rather strange.

Imagine you put an electron in a box. It stays there for the same reason that any object will stay in a box: there are walls in the way. If the particle hits the wall, the wall pushes back, just as a brick wall does if you wander absent-mindedly into it nose first. Let's keep it simple and say that the walls' repelling force is all or nothing: the electron feels nothing at all until it bumps into a wall, whereupon the repulsion is infinitely large. Then there's no way out.

This is the anodyne, staple model of introductory courses in quantum mechanics. It's not quite as arbitrary and artificial as it might sound, since it's a rough-and-ready way of describing any circumstances in which an electron might find itself confined to a restricted space: in an atom, say, or in an electronic transistor. But it's basically just a minimalistic way of getting the electron to stay put so we can solve its Schrödinger equation, deduce the resulting wavefunction, and see what this tells us about quantum behaviour.

Here's what the math tells us. The wavefunction's amplitude oscillates rather like a guitar string clamped at each end and plucked. It has particular frequencies of oscillation, selected by the condition that a whole number of peaks and troughs must fit precisely into the box. There's only a perfect fit for certain frequencies – or equivalently, for certain wavelengths of the wavefunction. And because the energy of the electron depends on the frequency with which its wave-like state oscillates – remember Planck's equation relating energy and frequency – there is a set of possible energy states rising progressively in energy like rungs of a ladder. In other words, the electron's energy is *quantized* as a result of its confinement in the box and the fact that it is described by the Schrödinger equation. An electron can only jump from one rung to another by gaining or losing energy of the right amount.

This isn't how a tennis ball in a box would behave. If the bottom is perfectly flat, the ball could be found with equal probability anywhere in the box: no location is preferable to any other. And the ball might simply rest in that position, with zero energy. Not so the electron. The lowest-energy state has a certain irreducible amount of energy – the electron is always 'moving', and in this state it is most likely to be found in the middle of the box, the probability decreasing the closer you get to a wall.

Here, then, is the quantum alternative to the mechanics of classical physics, as embodied in the equations of motion deduced by Isaac Newton in the seventeenth century. And how abstract and hard to visualize this description has become! Instead of particles and trajectories, we have

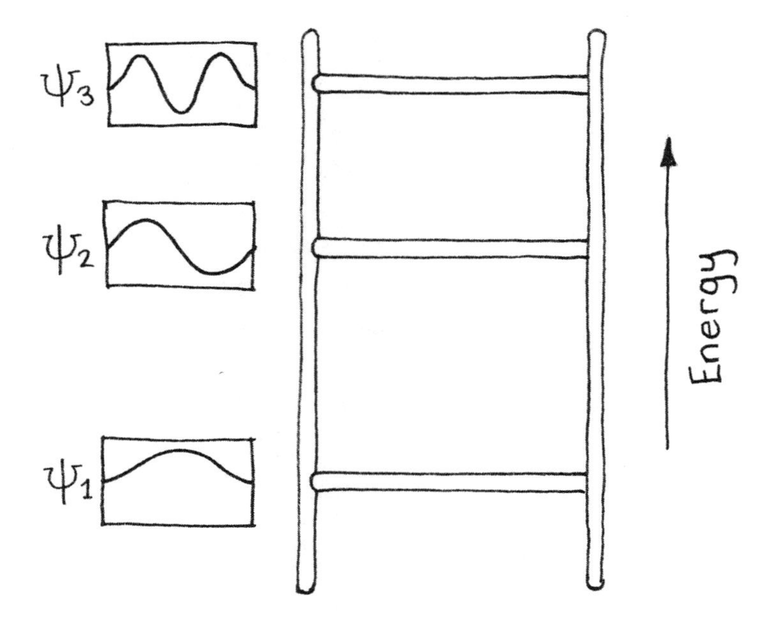

Wavefunctions Ψ and the corresponding 'rungs' on the energy ladder for the first three quantum states of a particle in a box. The amplitudes of all the wavefunctions are zero at the walls themselves.

wavefunctions. Instead of definite predictions, we have probabilities. Instead of stories, we have math.

It doesn't seem enough. What is the real nature of the electron that underlies these probabilities, this smooth, spread-out wavefunction?

Maybe we should picture the electron zipping around so fast that we can't easily see where it is, except that we can just about make out that it spends more time in some places than others. In this view, electrons confined in space – around the nucleus of an atom, say – are like a swarm of bees indistinctly glimpsed as they hover around a hive. At any instant each bee is *somewhere*, but it's only by making a measurement that you find out where.

But that's not the right way to think about a wavefunction – for it *says nothing about where the electron is*. I just told you, though, that the wavefunction says all we can know about the electron. If so, we have to accept that, as far as quantum mechanics (and therefore current science) is concerned, there simply *is* no 'where the electron is'.

OK then, let's accept that the electron doesn't *have* a location. It's not after all a concrete little particle, but is truly smeared, a kind of washed-out blot of electrical charge in the weft and warp of space. Will *this* image do instead? Can we think of the wavefunction as a description of a particle that is, at every instant, delocalized over space?

No, this image won't do either. For when it is measured, there it is: a point-like particle in a location as fixed, more or less, as the parking space where you left your car.

Both of these pictures – a particle blurred by rapid motion or a smear at every instant spread throughout space – testify to our determination to find *some* way of visualizing what the wavefunction is all about. This is perfectly natural, but that doesn't make the pictures correct. Born's probabilistic interpretation of the wavefunction reveals why quantum mechanics is so odd relative to other scientific theories. It seems to point in the wrong direction: not down towards the system we're studying, but up towards our experience of it. Here, then, is how we might express the reason why we can't use the wavefunction of an electron to deduce anything about what it is 'like' or what it 'does':

> The wavefunction is not a description of the entity we call an electron. It is a prescription for what to expect when we make measurements on that entity.

Not all quantum physicists would agree with that; as we'll see, some believe the wavefunction does refer directly to some deeper physical reality. But exactly what that belief implies is subtle, and is certainly unproven. Viewing the wavefunction as simply a mathematical tool for making predictions about measurement is a good default position, not least because it can save us from the error of inventing pictures of classical waves or particles in an attempt to envisage the quantum world. And this, at any rate, is what Niels Bohr and Werner Heisenberg thought; Heisenberg put it like this:

This doesn't mean the wavefunction tells us where the electron is likely to be at any instant, which we can then verify by making measurements. Rather, the wavefunction tells us nothing about where the electron is *until* we make a measurement. We can't even say what the electron 'looks like' until that measurement is made: it's not 'smeared-out charge', nor is it 'zipping about'. We should not, in truth, talk about the electron at all except in terms of the measurements we make on it. As we'll see, such linguistic rigour is all but impossible to sustain in practice – we are compelled, in the end, to talk about an electron that exists before we look. That's OK, so long as we recognize that we're then making an assumption outside of quantum mechanics.

•

Imagining the electron as a 'wavy particle' confined within a tiny box is a rather fruitful way of thinking about how atoms are constituted. One of the first successes of quantum theory was the model of the atom proposed by Bohr in 1913. It was adapted from an earlier picture suggested by the New Zealander Ernest Rutherford, in which he visualized these building blocks of matter as an extremely dense and positively charged central nucleus surrounded by negatively charged electrons. Rutherford and others refined this into a 'planetary model', where the electrons circulate in orbits like the planets around the Sun. They're confined not by walls around the edge of an atom but by the electrical force of attraction to the nucleus at the centre.

There was a big problem with the planetary model. It was known that charged particles going round in circles should radiate energy in the form of electromagnetic waves – that's to say, light. This means that the electrons should gradually shed energy and spiral into the nucleus. Atoms should rapidly collapse.

Building on Max Planck's quantum hypothesis in which energy was considered to be grainy rather than smooth, Bohr proposed that the electrons have quantized energies, so that they can't fritter it away gradually. They must remain in a fixed orbit unless kicked into another one with a different 'allowed' energy either by absorbing or radiating a quantum of light with the right amount of energy. Each orbit, Bohr argued, has only a finite capacity to accommodate electrons. So if all the orbits of lower energy than that of a given electron are already full, there's no way the electron can lose some of its energy and jump down into them.

This was a totally ad hoc picture. Bohr could offer no justification for why the orbits were quantized. But he wasn't claiming that this is what real atoms are like. He was simply saying that his model could explain the observed stability of atoms – and better, it could explain why atoms absorb and radiate light only at very specific frequencies. Louis de Broglie's wave picture of electrons later offered a qualitative explanation for why the Bohr atom had the properties it did. Electrons confined to particular orbits around the nucleus would have to have particular wavelengths – and thus frequencies and energies – so that a whole number of oscillations would fit long the orbital path, forming 'standing waves' rather like the waves in a skipping rope tied to a tree at one end and shaken (except that we can't answer the question 'waves of what?').

Knowing the nature of the electrical force that attracts an electron to the atomic nucleus, one can write down the Schrödinger equation for an atom's electrons and solve it to figure out the three-dimensional shapes of their

In a crude picture of Bohr's quantum atom, electron energies are fixed by the requirement that whole numbers of waves in their wavefunctions must fit around the orbits. Here the successive orbits accommodate two, three and four wave-like oscillations.

wavefunctions – and thus the probabilities of finding the electrons at any position in space. It turns out that the resulting wavefunctions don't correspond to electrons circulating the nucleus like planets, but have considerably more complicated shapes called orbitals. Some orbitals are diffuse spheres, perhaps with a concentric shell shape. Others have rather complex dumb-bell- or doughnut-shaped regions where the amplitude is large. These shapes can explain the geometries with which atoms join together in molecules.

•

The force of attraction that keeps an electron in the vicinity of a nucleus is not infinite, unlike the force exerted by the walls that confined our hypothetical electron-in-a-box. So electrons can be pulled off atoms, as they often are in chemical processes: movement of electrons, and their redistribution into new spatial patterns, lies at the heart of chemistry. What if the confining force of the walls for an electron-in-a-box is also less than infinite?

Then something odd transpires: we find that the wave-function of the electron in the box can penetrate *into* the walls. If the walls aren't too thick, the wavefunction can actually extend right through them, so that it still has a non-zero value on the outside.

What this tells you is that there is a small chance – equal to the amplitude of the wavefunction squared in that part of space – that if you make a measurement of where the electron is, you might find it *within* the wall, or even *outside* the wall. The electron can jump out, as though it has got *through* the wall. What's odd about it is that, according to the classical-physics picture, the electron simply doesn't have enough energy either to jump over the top of the walls or to punch a hole through them. Classically, it ought to remain in the box forever. But quantum mechanics tells us that, if we wait for long enough (or measure often enough), eventually the electron is bound to turn up outside.

This phenomenon is called quantum tunnelling. The electron (or any other quantum particle in such a situation) is said to be capable of tunnelling out of the box, even though from a classical perspective it lacks the energy needed to escape. Tunnelling is a real effect: it has been observed widely, for example in the way electrons get exchanged between molecules. There are experimental techniques and practical devices that rely on it. The scanning tunnelling microscope is an instrument that uses electron tunnelling between a sample and a very fine, electrically charged needle held just above it to produce atomic-resolution images of materials. Because the amount of electron tunnelling – and therefore

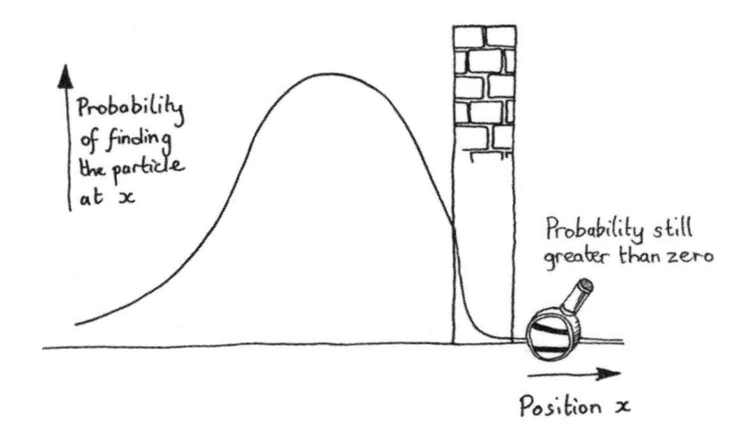

Wavefunctions can extend into and beyond walls with a less-than-infinite repulsive force and thickness, so that there is a chance of finding the particle on the other side of the wall - even though it hasn't enough energy to get 'through'.

the size of the electrical current passing between the needle and the sample – is very sensitive to the distance between them, the instrument can detect bumps on the sample surface only one atom high. The flash memories in hand-held electronics also work using electron tunnelling through thin layers of insulating material: a voltage is used to control the amount of electron tunnelling across the insulating barrier, so that information, encoded in electrical charge, can be written into and read out from the memory cells.

How should we think about tunnelling? It is often portrayed as another of those 'weird' quantum effects, a kind of magical act of vanishing and reappearing. But in fact it's not so hard to intuit – or at least, it is not so hard to *suppose*. So quantum particles can tunnel through barriers: well, why not? The feat is not possible within a classical picture, but it's *imaginable* if we don't worry too much about how it was achieved.

This doesn't mean, however, that we should picture the electron wriggling its way through the barrier. We can predict, using the Schrödinger equation, what we will measure in a tunnelling process, but we can't relate that to an underlying picture of an electron 'doing' anything. It's better to see this effect as a manifestation of the randomness that sits at the core of quantum mechanics. The wavefunction tells us where we might potentially find an electron when we look; but what we do find in any given experiment is random, and we can't meaningfully say why we find it *here* rather than *there*.

•

I don't expect you to give up that easily. Very well, you might say: suppose I accept that the wavefunction is nothing more than a formal device for letting us predict the likely outcomes of measurements. But the question then remains: *what is going on* to produce those outcomes?

Probably the most fundamental issue in quantum theory is whether or not this is a meaningful distinction to draw. Is there some 'element of reality' that the wavefunction represents, or is it just an encoding of accessible knowledge about a quantum system?

Some physicists argue that the wavefunction is a 'real' thing. Exactly what this means is often misrepresented, however. The wavefunction of an electron evidently does not correspond to some tangible substance or property, like an equation describing the density of air. For one thing, wavefunctions generally contain 'imaginary' numbers – ones involving the square root of –1, which is not something that has a physical meaning.

But when scientists refer to the wavefunction as real, what they mean is that there is a unique, one-to-one relationship between the mathematical wavefunction and the underlying reality it describes.

Wait a moment! Didn't I cast doubt on the idea that we can speak about an 'underlying reality' to quantum mechanics at all? Yes I did – and that's why all suggestions that the wavefunction is 'real' are predicated on the assumption that there *is* after all some deeper picture in which particles have concrete, objective properties regardless of whether or not we measure them (or even *can* measure them). This picture is commonly called a realist view. There is no reason to think that it *is* a valid way to think about the world, and a fair bit of evidence implying that it is not. Yet some scientists continue to feel deep down that realism – an objective world 'out there' – is ultimately the only option that makes sense.

The concept of 'reality of the wavefunction' then asserts that the mathematical wavefunction can be directly and uniquely related to this objective reality: it refers to real and unique 'things' – ball-bearing particles if you will – and not just to our imperfect state of knowledge about them. Some experiments have suggested that *if* the realist view is valid, then the wavefunction must indeed be 'real' in this sense.

This picture of quantum mechanics is said to be *ontic*: from 'ontology', meaning the nature of things that exist. The alternative view is that the wavefunction is *epistemic*: as Heisenberg asserted, it refers only to our state of knowledge about a system, and not to its fundamental nature (if such a concept has any meaning). In this latter view, if a wavefunction changes because of something we do to the quantum system, it doesn't imply that the system itself has changed, but only that our knowledge of it has.

Actually even Heisenberg's formulation doesn't go far enough, because referring to a 'state of knowledge' seems to imply some underlying facts to which we have imperfect access. Better to say: in the epistemic view, the wavefunction tells us what expectations to have about the outcomes of observations or measurements.

This distinction between ontic and epistemic viewpoints is the Big Divide for interpretations of quantum mechanics. It's where you must reveal your true colours. Does the wavefunction express a limitation on what can be known about reality, or is it the only meaningful definition of reality at all?

Definitions of reality are immensely subtle philosophical fare; but if we accept some physicists' view that quantum reality begins with the wavefunction, then we can never adduce a reason why, when we make a measurement, it gives the result we observe. That would make quantum mechanics unlike any scientific idea previously encountered. As the quantum physicist Anton Zeilinger has put it, the theory may expose 'a fundamental limitation of the program of modern science to arrive at a description of the world in every detail'.

This possibility seemed to Einstein to be a profoundly *anti-scientific* idea, because it meant relinquishing not just a complete description of reality but the notion of causality itself. Things happen, and we can say how likely they are to happen, but we cannot say why they happened just as or when they did.

Take radioactive decay. Some radioactive atoms will decay by emitting an electron from *inside* the nucleus: this electron is, for historical reasons, called a beta particle, but it's just a common-or-garden electron. Atomic nuclei don't exactly contain electrons – we saw that these orbit outside the nucleus. But they do contain particles called neutrons, which may spontaneously decay into an electron, which gets spat out, and a proton, which stays in the nucleus.* Beta decay of carbon-14, one of the natural forms of carbon atoms, is the process used for radiocarbon dating, and it transforms the carbon atom to a nitrogen atom.

Beta decay is a quantum process, so the probability of the neutron decaying is described by a wavefunction. (It's actually a kind of quantum tunnelling process: the electron tunnels out of the nucleus, where it would otherwise be bound by electrical attraction.) All the wavefunction can tell you is the probability that the decay will occur – not *when* that will happen. Take any specific atom of carbon-14, and its decay could happen tomorrow, or in 1,000 years. And there's nothing, *nothing*, you can do to figure out which it will be, for all the carbon-14 atoms look alike.

However, once you know the probability of beta decay, you can estimate when, from a sample of a billion or so atoms, exactly half of them will have decayed. This is just a matter of averages. By the same token, if by some strange coincidence you're in an antenatal class with ten other expectant mothers with the same due date, you can't be sure exactly when any one of the babies will be born but you can make a pretty good estimate of the date by which 50% of them are likely to have been born. The bigger the sample, the better the estimate. For radioactivity, this time taken for half of the atoms in a sample to decay depends on the detailed specifics of the type of nucleus in question, and is called the half-life.

* Beta decay also produces a particle called a neutrino, which carries off some energy and a tiny amount of mass.

For carbon-14, the half-life is 5,730 years, which is just right for estimating ages of objects derived from living things over the past several centuries or millennia.

What makes the quantum situation of radioactive decay any different from this classical (and, forgive me, rather unfortunate) analogy of childbirth? We have every reason to suppose that, if we were able to monitor the biology of each pregnant mother and baby sufficiently closely, we could understand exactly why the respective birth process started when it did – perhaps some hormonal threshold was reached, say. But for radioactive decay, there is nothing you can monitor to explain why a particular atom decayed when it did. There is nothing we can call a reason.

OK, so atomic nuclei are pretty hard to peer into. But that's not the root of the problem. It's that we simply can't, for quantum processes, talk about a historical progression of events that led to a given outcome. There's no story of how it 'got' to be that way.

But – and this is the perplexing thing about quantum mechanics – it often seems as though you can tell a perfectly rational and convincing story of just this sort! You can 'fire' a photon from a laser at some initial time, and then at some later time you are highly likely to detect it at another position just as though it went there along a straight-line path from the laser at the speed of light. It seems the 'reason' you detected it at B is that it left A and reached B along the most direct path.

What's wrong with that tidy story of cause and effect? Sometimes there really is no harm in telling it as if it happens that way. But we must try as hard as we can to keep that 'as if' in sight. For in some situations such a narrative won't work at all.

Quantum particles aren't
(but sometimes they

in two states at once
might as well be)

The question now hurtling towards us might sound like pedantic, navel-gazing philosophy, but really there is no escaping it:

What do we mean by 'is'?

Is an electron a particle or a wave? It can, in different circumstances, display the characteristics of either – or even a bit of both. But as for what an electron 'is', all we can talk about for sure is what we can see and measure, not what causes those observations. We must say that wave–particle duality is not a property of quantum objects but a feature often invoked (to questionable benefit) in our descriptions of them. They don't have 'split personalities'.

The same applies to the much-vaunted notion that quantum particles can be in two places at once – or more generally, in two states at once. This, too, is not really true, but again I wouldn't go so far as to say it is wrong. We are suspended in language. From the human perspective, it certainly looks as though quantum objects can possess two different, even contradictory, values of some property at once. But the human perspective is the wrong one for understanding quantum mechanics. Nonetheless, it's all we have.

Don't despair. We might not have the right cognitive or linguistic tools, but at least we know, more clearly than Einstein and Bohr could have done, what we're missing.

Yet what a horrible word 'state' is anyway: cold and formal while at the same time vague and deceptively pedestrian. We seem compelled to use it without quite knowing what we're talking about. In science, the 'state' of an object usually has a trivial meaning: it refers to some or all properties of the object. My state right now is rather overheated (summer has finally arrived) and in need of a cup of tea. The state of my desk can be a little more precisely defined: it is, among other things, fairly rigid, with a temperature of around 20°C, and is faux-wood tawny in colour. A state tells us something about how things are. And as you'll probably now appreciate, that's why it is a difficult concept in quantum mechanics – because quantum mechanics *doesn't* obviously tell us about 'how things are'.

By the 'state' of a particle, we mean the collection of properties that in some sense label it for us. (I'm being deliberately vague with 'in some sense', and 'for us' hides some difficult questions too.) *This* atom is not *that* atom because it is *here* and not *there*,* but also because it is travelling at *this* speed, and because its electrons have *these* energies, and so on.

The classical idea of a state generally has an exclusive aspect to it. Macroscopic objects can be a bit of this and a bit of that – a bit rigid but somewhat flexible, or kind of reddish brown. But they can't be in mutually exclusive states: *here* and *there*, having a mass of 1 g and also of 1 kg. I can't be cycling at 20 mph at the same time as cycling at 10 mph. And my cycling jacket can't be bright yellow at the same time as being pink. It can be a mixture of both, but it can't be all yellow and all pink. This seems common sense.

So it's understandable that, when we hear that quantum particles can be in more than one state at the same time, we

* The distinguishability of quantum particles is in fact an extremely important issue, but there's no need to delve into it here.

struggle to see what that could mean, and we start to talk about quantum weirdness – or figure that we're too plain dumb to comprehend quantum mechanics. Perhaps we can manage the idea of a particle being in more than one *position* at a time, if we think of it as being kind of smeared out or blurry, like a gas. This, as I have explained, is *not* the best way to think about such an object – but all the same, it's a mental picture we can cling to. Yet to claim that, let's say, a particle may have two different velocities at the same time: that seems not just senseless but inconceivable.

But again, talking about the 'two-states-at-onceness' of a quantum particle in these terms isn't strictly proper at all. For a start, a quantum state, as defined by a wavefunction, encodes the expected outcomes of measurements of specific observable properties. So what we mean in such a case is that we can create quantum states with wavefunctions so that, if we do an experiment to measure a property of the particle, we might observe either of the two outcomes. But what then is actually going on for the particle – what, you might say, is its 'isness' – both *before* and *after* we make the measurement? The various interpretations of quantum mechanics are largely differentiated by their answers to these questions.

•

This 'two (or more) states at once' is called a superposition. The terminology conjures up the image of a ghostly double exposure. But strictly speaking a superposition should be considered only as an abstract mathematical thing. The expression comes from wave mechanics: we can write the equation for a wave as the sum of equations for two or more other waves.

Here's another way of saying that. The wavefunction is a *solution* to the Schrödinger equation, much as $x = 2$ is a solution to the equation $x^2 = 4$. The wavefunction is an expression that makes the 'equals' sign in the Schrödinger equation true.* In general there is not just one of these solutions; there are many, just as another solution to $x^2 = 4$ is $x = -2$. That's why there's a whole bunch of energy states for an electron in a box, or in an atom.

Superpositions arise because, if two wavefunctions – let's write them as Φ_1 and Φ_2 – are solutions to the equation, then so is any simple combination of these two, such as $\Phi_1 + \Phi_2$. A sum of the two wavefunctions does seem to invite the notion that they are in some sense 'superimposed', but we have to be careful. An equally valid combination is $\Phi_1 - \Phi_2$, and how are we supposed to interpret that?

By a 'simple' combination here, I mean what mathematicians call a linear combination: roughly speaking, that means a sum like one wavefunction plus or minus the other. It excludes more complex combinations involving things like higher powers of the wavefunctions, such as $\Phi_1^2 + \Phi_2^3$. The fact that these linear combinations or superpositions are permissible states of the system in Schrödinger's quantum mechanics has nothing to do with its being quantum – it follows from the fact that it's based on wave physics. Superpositions of waves are just other waves. Superpositions of quantum states only seem

* Let me give you a glimpse of the Schrödinger equation, since it is not so fearsome, at least to look at. One version simply reads $H\Psi = E\Psi$, where Ψ is the wavefunction and E is the system's energy. H is a term called the Hamiltonian operator, and contains the various factors that affect and determine the system's energy.

odd because the wavefunctions are used to describe the properties of entities that we can also regard as particles – meaning that such particles seem to be able to have two or more values of their properties at once.

So what's the right way to think about superpositions of quantum states? Let's consider a single photon, a quantum of light. Light, as I explained earlier, is an electromagnetic field: an oscillation of an electric field coupled to an oscillation of a magnetic field. These up–down vibrations of the fields have particular orientations in space, rather like the up–down oscillation of a piece of rope tied to a post and shaken. This orientation is called polarization. A polarizing filter – like those that reduce glare in sunglasses and cameras – is a material that will allow only photons of a particular orientation to pass through. So a photon's state includes some value of its polarization, defined relative to a particular direction in space. But photons can also be created in superpositions of polarization states: say, an up–down vertical polarization combined with a side-to-side horizontal polarization.

What does this superposition of photon states look like? We typically speak of it as a kind of mixture of the two polarization states (although strictly speaking a 'mixture' has a different, technical meaning in quantum theory). Does this mean that sometimes the photon is oscillating vertically and sometimes horizontally? Not really. Does it mean that half of the photon is vertically polarized and half is horizontally polarized? That doesn't have an obvious meaning at all. Then what?

Niels Bohr's answer was simple: don't ask. The wavefunction of superposed states doesn't say anything about what the photon is 'like'. It is a tool for letting you predict what you will measure. And what you will measure for a superposed state like this is that sometimes the measurement device registers a photon with a vertical polarization, and sometimes with a horizontal one. If the superposed state is described by a wavefunction that has an equal weighting of the vertical and horizontal wavefunctions, then 50% of your measurements will give the result 'vertical' and 50% will indicate 'horizontal'.

If you accept Bohr's rigour/complacency (delete to taste), we don't need to worry what the superposed state 'is' before making a measurement, but can just accept that such a state will sometimes give us one result and sometimes another, with a probability defined by the weightings of the superposed wavefunctions in the Schrödinger equation. It all adds up to a consistent picture.

But it's not a picture we can visualize in terms of particles doing stuff, or even of quantum fields vibrating. Is there an experiment that can help us think about what the particles are up to? Yes there is – but what it shows is just how perplexing any attempt to pin down 'what really happens' in a quantum system is.

•

It is arguably *the* central experiment in quantum mechanics. And no one truly understands it.

The so-called quantum double-slit experiment is delightfully simple to explain. It's also simple to see what the results are. What we don't understand is how to interpret those outcomes in terms of underlying processes: in terms of things doing stuff.

The experiment exploits a characteristic phenomenon of waves called diffraction, which is a consequence of how waves interfere. If two waves encounter one another, they can enhance or attenuate the oscillations depending on the relative timings of the train of peaks and troughs. The total amplitude of two overlapping waves is simply the sum of their individual amplitudes. So when two identical wave crests meet and overlap, they produce a crest of twice the amplitude. But if a crest coincides with a trough, they cancel out and the amplitude is zero. Their sum can also be anything in between, say if a crest of one wave coincides with another halfway between peak and trough. The stage of a wave in this cycle of crests and troughs is called its phase. Thus, waves that overlap and interfere when they are in phase with one another – when the peaks and troughs are in step – reinforce each other (they are said to interfere constructively) whereas out-of-phase waves annihilate one another (they interfere destructively). For two interfering light waves, constructive interference will increase the brightness whereas destructive interference will produce darkness.

Imagine creating two wave sources by passing a single train of waves through two small slit-like gaps in a wall, spaced close together. As the waves pass through the slits, they will radiate on the far side like ripples from a stone dropped into a pond. Where these ripples overlap, there

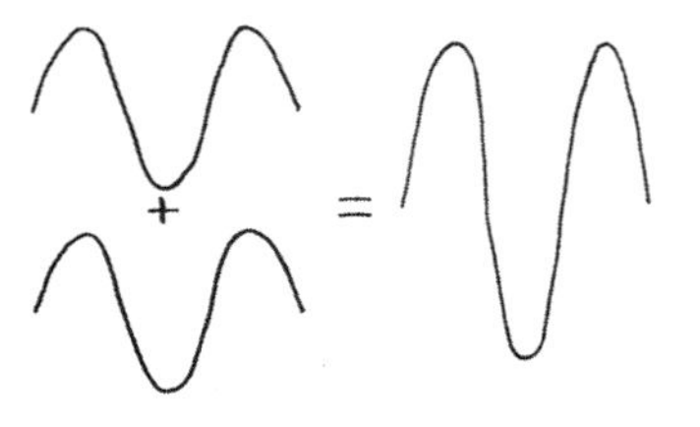

Constructive interference (bright)

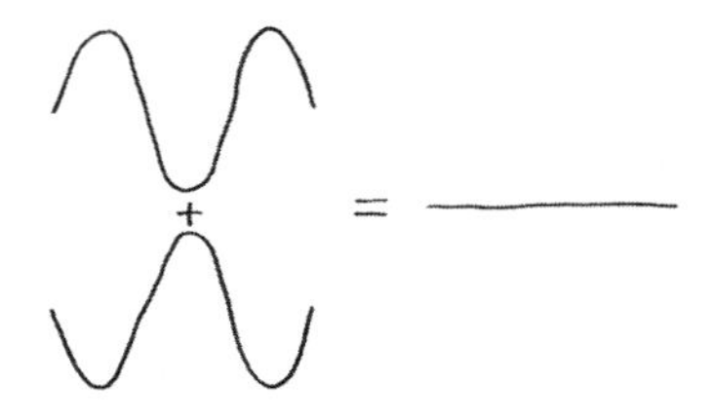

Destructive interference (dark)

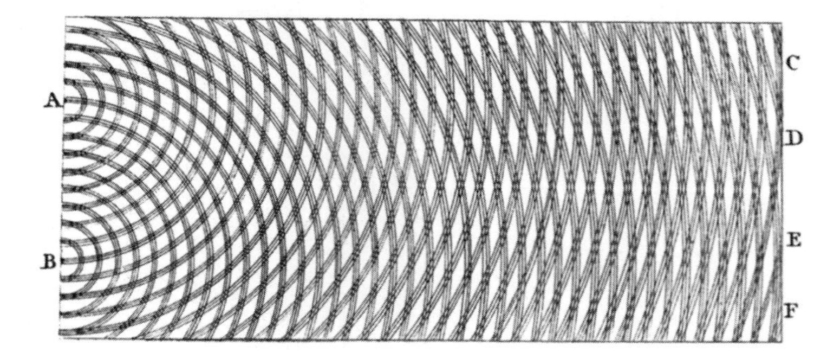

The English scientist Thomas Young first explained the diffraction of light passing through double slits in the early 1800s. This is his drawing of interference fringes – the dark bands are C–F – produced when light passes through the slits A and B. Young presented it to the Royal Society in London in 1803.

is a regular pattern of constructive and destructive interference. If these are light waves, then a screen on the far side of the slits receives a stripy pattern of light and dark bands, called interference fringes. This is an example of diffraction: the spreading and interference of light passing through gaps or bouncing off arrays of objects.

All this was understood by the early nineteenth century. The interference pattern is strictly a wave phenomenon. Compare it with what we'd expect if we fired particles at the double slit instead – using a sandblaster, for example. Now all we'd see on the screen is an image of the two slits picked out by particle impacts on the screen: the slits are simply acting as a kind of patterning mask.

But what if Louis de Broglie was right that quantum particles exhibit wave-like properties? Then we might expect to see interference fringes for the particles. And we do.

The interference and diffraction of quantum 'particles' was first observed between 1923 and 1927 by the physicists Clinton Davisson and Lester Germer, working at Bell Labs

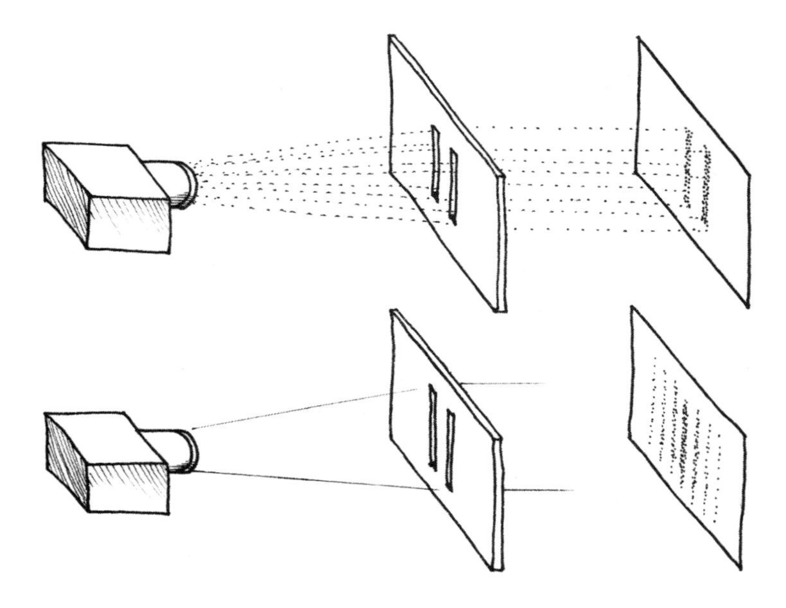

A double-slit experiment with classical particles just produces a projection of the two slits in the particle impacts on a screen (top). For quantum particles, however, passage through the double slits produces a series of bands in which there are many impacts, separated by gaps with virtually none: wave-like interference fringes (bottom).

in New Jersey. They looked for wave-like interference from beams of electrons emitted from a hot metal electrode and accelerated by an electric field. Davisson and Germer didn't actually use double slits, though. Instead they looked at the kind of interference that occurs when waves bounce off regular arrays of objects spaced at a distance similar to the wavelength of the waves. The waves that bounce off different elements of the array can interfere with one another, again producing 'light' and 'dark' regions.

According to de Broglie's proposal, electrons in a beam produced this way have wavelengths similar to the spacing between atoms in the crystal lattices of metals. Davisson and Germer found that indeed the electrons were diffracted when fired at a piece of nickel. The English physicist George Paget Thomson also demonstrated this effect at much the same time. Davisson and Thomson shared the 1937 Nobel Prize in Physics for verifying de Broglie's bold thesis.* (De Broglie himself had taken the 1929 prize.)

The Davisson–Germer experiment is often cited as a demonstration of the wave-particle duality of electrons. As we saw, this isn't a particularly helpful expression. The double-slit experiment reveals why.

•

If we conduct a double-slit experiment for electrons, we see interference fringes. We can, for example, place a phosphor screen on the far side that reveals the arrival of an electron as a bright, glowing spot: the process used in old-fashioned 'cathode-ray tube' television screens. You get the same result for photons of light, but I'll talk about electrons here because we're more used to thinking of them as particles, with mass and all.

Suppose now we make the electron beam so faint that on average only one electron passes through the slits at a time. Each particle leaves the gun one at a time, and each hits the screen one at a time, and only then does the next electron take flight. So there are now no bright and dark interference fringes on the screen, corresponding

* Why not Germer too? He was the junior partner in the team, and in those days that meant you could not expect a share of the glory. Reputedly good-natured, he seems to have harboured no ill feelings.

to intense or dim parts of the electron beam. There are just single flashes each time an electron hits. No longer waves, but particles – right?

Let's see. As the experiment progresses, we keep a record of where the electrons are hitting. And here's the surprise. The electrons are detected one particle at a time – but over time, the pattern that builds up of where they are striking proves to be a series of parallel bands in which a high density of hits alternates with a low density. This isn't the simple 'shadow' of two slits that we'd expect from the same experiment with a feeble sand-blaster. These are unmistakably interference fringes.

We can't explain this result in terms of particles, but only in terms of 'electron waves'. And whereas we might have been content enough to believe that electrons in a

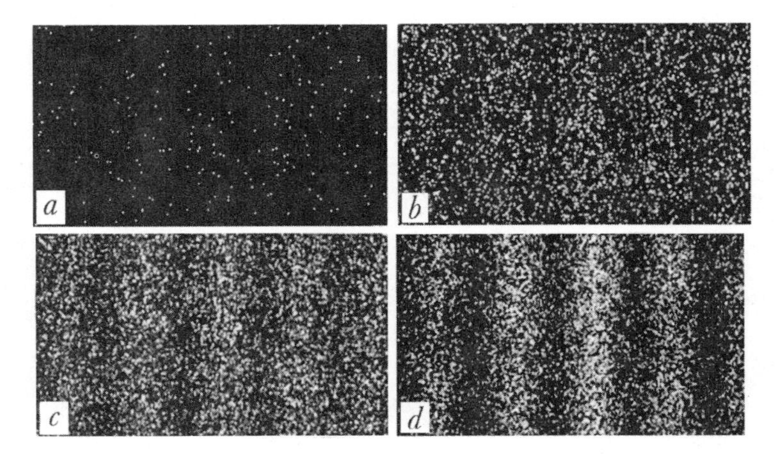

As particle impacts on a screen gradually accumulate (from *a* to *d*) in a double-slit experiment with a weak electron beam, what at first looks like random impacts (*a*) becomes revealed as bright and dark interference fringes (*d*). These are the results of an actual experiment performed in 1987 by the Japanese physicist Akira Tonomura and his collaborators.

bright beam are wave-like and can be diffracted by the double slits, it is hard to understand how one-by-one passage of what seem to be particles (judging from the discrete bright spots that appear on the screen) can produce wave-like interference. We're forced to conclude that 'wave-like' electrons can interfere with *themselves*.

But that requires us to believe that each individual electron passes through *both slits* – for there must be two sources of the electron waves on the far side if there's to be interference. What's going on? Why should the electron act like a particle before and after encountering the slits, but become a spread-out wave as it passes through them?

No, that can't be the right way to look at it. So let's get smarter. If we can detect a particle-like electron, pinpointed in space, both before and after the slits, why not try to do that within the slits itself? Why not put a detector behind one slit that can report if an electron went through it? Let's use a device that can sense the passage of an electron without actually capturing it. If the detector at one slit fails to detect an electron, yet there is a bright flash on the screen owing to an electron impact, we know that the electron must have gone through the other slit.

It's possible to set up experiments like this to measure the routes of quantum particles like electrons, photons or atoms. And we can indeed detect whether a particle goes through one slit or the other.

The problem is that, when we do, the interference fringes disappear. Instead, we just see the double-slit acting as a mask, generating two bright bands on the screen. Now the particles really are particle-like, and we're no longer confronted by the puzzle of a 'particle' passing through both slits at once.

Suppose now we turn off the electron detector. We haven't done anything to the slits, nor to the electrons passing through them, except that we're no longer detecting their path. And yet this decision to cease observing the electrons makes the interference fringes reappear.

Truly, this is what happens. The experiment has been done countless times.

Are electrons perverse? So long as we don't try to figure out which slit they go through, they will behave as if they go through both at once. But if we try to pin down which slit they pass through, they only go through one. The mere act of making the measurement – even if we can be pretty sure that the measurement shouldn't obstruct or influence the electron's path – appears to turn a wave into a particle.

Yes, appears to. *Does* the electron really pass through both slits at once when we're not looking at its path? *Does* it change from wave to particle when we do look? These are, according to Bohr's view of quantum mechanics, illegitimate questions, precisely because they are insisting on some microscopic description underlying the measurements we make. Bohr argued that there is nothing in quantum mechanics that permits us to formulate such a description. That is not what the Schrödinger equation is about. It just predicts the outcomes of measurements.

And if we use quantum theory to calculate what we should see in the double-slit experiment with and without monitoring the passage of the particles through specific slits – yes, we can do that for a variety of possible schemes for deducing the paths taken – then the theory predicts just what I've described. And this is because, when we *don't* look but not when we *do*, the electron's wavefunction can be written as a linear combination of wavefunctions for electrons passing through each slit: a superposition of two 'paths'.

If we try to imagine a scenario involving particles and waves that can give rise to these observations, we will get stuck, because we are then faced with the prospect of waves that somehow magically sense that they are being observed and so decide to become particles instead. If, however, we simply use Schrödinger's quantum mechanics to describe these experiments, the equations predict the right outcomes.

So, said Bohr, we had better just stop there. As he put it:

> There is no quantum world. There is only an abstract quantum physical description. It is wrong to think that the task of physics is to find out how nature *is*. Physics concerns what we can *say* about nature.

The is the central tenet of the so-called Copenhagen Interpretation of quantum mechanics, developed by Bohr and his colleagues in the Danish capital during the mid-1920s.* It's an interpretation that doesn't so much tell us 'what is happening', but rather, proscribes what we can legitimately ask about it.

At face value that sounds crazy. Why make a mathematical theory like quantum mechanics if you don't think

* Strictly speaking, one should not refer to the 'Copenhagen Interpretation' as something fixed and monolithic. As with some other interpretations of quantum mechanics, the Copenhagen position was expressed by different proponents in different ways: Bohr's view was not identical to Heisenberg's, and so on. This personal inflection of the Copenhagen Interpretation applies also to those who prefer it today. The statements I shall make here about the 'Copenhagen Interpretation', however, generally refer to a shared core of ideas.

it can tell you about the system it is supposed to describe? But Bohr argued that quantum theory tells you something more meaningful – indeed, the only thing that *can* be meaningful. It tells you what you will find when you try to *investigate* that system. It tells you about measurements.

•

This just doesn't seem to be a complete and satisfying story, does it? It feels as though we *should* be able to speak of the electrons taking particular paths between leaving the electron gun and striking the screen.

This is an instinct deeply ingrained by experience. If we see an aeroplane go into a cloud and come out the other side, it is plainly absurd to doubt that it followed some particular path during the time we couldn't see it.

But on the scale of electrons and photons, the notion of trajectories starts to break down. Curiously, this might be easier to accept if it broke down utterly: if there was no telling where an electron might turn up and we just had to shrug our shoulders about how it got there. But we *can* measure the paths of such objects. If we place a detector at any point between the source and the screen, we'll find our intuition confirmed: the electrons seem on the whole to be taking straight-line paths, provided that no objects get in the way to scatter them in other directions. However, the moment we stop making such measurements and leave the particles up to their own devices, they can behave in ways that make this notion of paths nonsensical – for example, by forcing us to say that they 'pass through both slits at once'.

The implication seems to be that there is something strange about the act of measurement itself.

Let me add a final word of warning. In a formulation of quantum theory called quantum electrodynamics, developed in the 1950s and 60s by Richard Feynman together with Julian Schwinger and Sin-Itiro Tomonaga, the path that a quantum particle takes as it travels through space takes into account not just straight-line trajectories but *every route possible*. That's to say, the equations of quantum electrodynamics contain terms that correspond to every path, however tortuous and crazy. However, when you add all these terms up, most of them cancel out – the wavefunction has essentially zero amplitude throughout most of space. So it's sometimes said that quantum electrodynamics really does show that an electron or a photon goes through both slits in the double-slit experiment – because it takes every path 'at once'.

However, this picture is just a *metaphor for the mathematics*. You can think of the particle taking all possible paths if you like, but you can never show that they do. To interpret quantum electrodynamics this way is to attempt to tell a classical story about quantum mechanics. *The electron or photon does not take all possible paths.* To imagine that it does is not just mistaken; it is fundamentally the wrong way to think about quantum mechanics.

So what's the right way? Now you're asking.

What ‘happens’ depends on

what we find out about it

Everything that seems strange about quantum mechanics comes down to measurement.

If we take a look, the quantum system behaves one way. If we don't, the system does something else. What's more, different ways of looking can elicit *apparently mutually contradictory* answers. If we look at a system one way, we see *this*; but if we look at the same system another way, we see not merely *that* but *not this*. The object went through one slit; no, it went through both.

How can that be? How can 'the way nature behaves' depend on how – or if – we choose to observe it?

•

In the early days of this new physics, the problem was often debated in terms of the 'role of the observer'. That the observer *had* a role at all was deeply troubling, because it seemed to challenge the very concept of science. If what we see depends on what questions we ask, whither then the idea of an objective world, governed by rules that pertain independently of our attempts to figure them out? As Heisenberg put it, science had ceased to be a way of peeking unnoticed at the world, and instead had become 'an actor in [the] interplay between man and nature'.

But that seemed to make scientific results contingent on the circumstances of their observation. Surely the whole point of a scientific experiment is to provide knowledge that can be generalized beyond the particular conditions under which it was obtained? Otherwise, what's the point? If I (and a team of thousands) smash two protons together in the Large Hadron Collider at CERN and I see a new particle, I want to be able to conclude more than that I have discovered a new particle that appears when the LHC smashes protons together (and which I'd otherwise be obliged to call something like the 'LHC-smashon'). I want to be able to assume that the new particle is a feature of nature, not of the specific experiment that made it. If experiments couldn't answer any questions beyond those relating to that experiment alone, science would be nigh on impossible.

You might feel, correctly, that there's nothing intrinsically surprising or extraordinary about the act of observation influencing the outcome. It's especially common in the behavioural sciences. For example, suppose we're trying to figure out how honest people are when they are playing cards. We say that one player has to leave the room for a moment, and she puts her cards face down on the table. Will her opponent look at them? Of course I won't, says everyone. So we conduct that experiment in the lab – and sure enough all players are scrupulously honest. Yet when we monitor this situation for games played in ordinary locations (where we can't keep a close watch on people's actions), we find clear statistical evidence that some players must have cheated.

Obviously, players are altering their behaviour when they know (or suspect) they are being observed. There's no mystery here, no threat to the idea of an objective reality independent of observation. We just have to get smarter about doing the observing, so that we can eliminate this observer effect. It's a procedural problem.

People, however, know (or at least suspect) when they're being observed. But electrons and photons don't! It's not hard, though, to imagine how similar observer effects could arise even for non-sentient systems. Imagine that you have a solution of some chemical that kills bacteria – but that it *doesn't* work as a bactericide when you check that the chemical is present in the solution before administering it, using the technique of spectroscopy. It only works when you don't look. Weird? Not really. Spectroscopy involves shining a laser beam through the solution. So it could be that the laser disturbs the solution in some way: the light, as well as probing the molecules present, might actually break some of them apart, say. Then the very act of ensuring that the molecules are present destroys them.

Is there some analogous physical effect that the act of observation has on a quantum system, which alters its properties and behaviour?

It's extremely hard to see quite how it could work – for any such effect doesn't seem to depend on exactly how the observations are made. You could, for example, use several different detection methods to figure out which slit an electron or photon passed through in a double-slit experiment, but the result would always be the same: the interference would vanish. It seems to be the *fact* of detection, not the *method*, that makes the difference. It's not easy to see how any physical theory working at the level of known interactions between particles can account for that.

According to the Copenhagen Interpretation, this 'observer effect' is precisely what we should expect, given the mathematical structure of quantum mechanics. It's only strange if we insist on asking about physical causes rather than just predicting results. But quantum mechanics (said Bohr) can make no claim to tell us about such causes.

This is commonly called an instrumentalist view: crudely, it says quantum theory offers only prescriptions, not descriptions. To many researchers, it feels defeatist and disheartening. If I make a spectroscopic measurement using a laser to probe molecules in solution, I expect to be able to say something about those molecules. It would seem pretty pointless if all my theory could do was to say 'The laser light will get dimmer at the wavelengths of green light' while prohibiting me from concluding anything about the molecular processes responsible. Why even then make the measurement? No, surely we need to be able to talk about the connection between our experience and an underlying reality?

•

The relationship between what we observe and what *is* has long preoccupied philosophers. In the eighteenth century David Hume argued that we can never be certain about interpreting causation. If we find that A seems invariably to be followed by B, we might *infer* that A causes B, but that inference can't be proved correct. In *Critique of Pure Reason* (1781), Immanuel Kant went further, saying that we have no access to the world that is not mediated by experience. He called the world as it 'is' the *noumenal* world or the *Ding an sich*: the 'thing in itself'. But all we can know is the *phenomenal* world: that which is registered by the senses and the mind's tools of understanding. This holds our conception of the world hostage to fallible powers of perception and reasoning. If we become capable of reasoning more precisely, the phenomenal world changes. Most scientists feel instinctively that experience and consciousness should be a secondary phenomenon, a mere mediator rather than the primary ingredient for cooking up a concept of what reality could mean. But some philosophers, notably the Phenomenologists starting with Edmund Husserl (and anticipated by William James), have attempted to do so. A few physicists who think about the interpretation of quantum mechanics are now taking an interest in their ideas.

Today most scientists would accept that our reliance on sensory data puts us at one remove from any *Ding an sich*: all our minds can do is to use those data to construct its own image of the world, which is inevitably an approximation and idealization of what is really 'out there'. Stephen Hawking has written that 'mental concepts are the only reality we can know. There is no model-independent test of reality.'

This, however, is no big concession. Scientists tend to deal with it, often unconsciously, by cleaving to what philosophers call naive realism: assuming that we can accept at face value what our senses, with all their limits and flaws, tell us about the objective world 'out there'. Bohr, influenced by Kant's ideas, went further. He said that the world revealed by experience – which is to say, by measurements – is the *only* reality worthy of the name.

That might seem like metaphysical juggling. If we can't access anything beyond what experience shows, does it make any difference whether we choose to regard the deeper layer as 'real' or not? But the Copenhagen Interpretation claims that the act of measurement actively *constructs* the reality that is measured. We must abandon the notion of an objective, pre-existing reality and accept that measurement and observation bring specific realities into being from a palette of possibility. As Bohr's young colleague and fellow Copenhagenist Pascual Jordan put it, 'observations not only disturb what has to be measured, they produce it ... We compel [a quantum particle] to assume a definite position.' In other words, Jordan said, 'we ourselves produce the results of measurements'.

That's really radical. Some would (and did) say it is heretical.

•

The 'measurement problem' is another of the commonly misunderstood notions in quantum physics. It's often interpreted as meaning that we can't investigate anything without disturbing it, and that as a result science becomes wholly subjective. Neither clause is accurate.

Almost all of science is utterly untouched by the quantum measurement problem, and remains to all meaningful purposes an objective investigation of a world 'out there'. Even at the atomic scale we can generally make measurements without fear that we are significantly disturbing, let alone that we are determining, what we see. The disturbances are usually so small as to be insignificant. When, for instance, we measure in a lab the strength of a new material, we can obtain a value that is an intrinsic and reliable property of that material, useful for predicting how well it will perform in a building or a bone implant. We don't influence the outcome of the experiment by our choice of how to conduct it (not, at least, if the experiment is well designed). Any small disturbance to the system's properties that our intervention makes can be estimated and understood.

Besides, the idea that the quantum measurement problem is a matter of 'disturbing' what is measured is *exactly what the Copenhagen Interpretation denies*. That picture is predicated on the assumption that the system we are investigating has a particular property or character, and then in we blunder and change it with our clumsy measurement. The Copenhagen Interpretation, in contrast, insists that the system has no particular property or character *until* we make the measurement. In an extreme view, this implies that there is no such thing as the 'system' at all until we make the measurement.

The corollary is that different measurements produce different realities. Not just different results, but different realities – and what's more, ones that are not necessarily compatible with one another. This is why discussions of the interpretation of quantum theory often invoke 'paradoxes' or inconsistencies. The word gets overused; sometimes the 'paradox' isn't really a logical contradiction but just something that is hard to explain or understand. All the same, such 'paradoxes' have an important role in illustrating why quantum mechanics confounds intuition. They generally arrange quantum outcomes in such a way as to apparently permit the answers Yes and No simultaneously. Whatever we are to make of that, we must surely aspire to do better than shrug and call it 'weird'.

•

Bohr's allegedly instrumentalist view is often misrepresented. By denying that we can relate the predictions of quantum theory to an underlying stratum composed of interactions between objects (or at least, composed of *something*), he wasn't exactly denying that any such stratum exists. He was proposing that we need a new view of what 'quantum reality' can mean.

The conventional view is that a scientific experiment investigates and illuminates the phenomena that produce the result. Often in the physical and biological sciences we make observations on a macroscopic scale that we try to understand in terms of processes at smaller scales: how atoms, molecules or cells are moving and interacting. And this is a valid and productive way to conduct science. We can meaningfully say that my coffee cup and the view out of my window are – and indeed I am – somehow generated by processes and effects operating at smaller scales. This hierarchy is sequential: the properties and principles at one scale *emerge* from those operating at the level below. The solidity, brittleness and opacity of the coffee cup can be understood in terms of the atoms and molecules that make up its fabric, congregated in vast numbers.

But quantum mechanics disturbs this hierarchy. In Bohr's view, quantum experiments like the double slits can't be considered in terms of macroscopic outcomes resulting from underlying microscopic processes. We have to regard the macroscopic process itself as an irreducible phenomenon, inexplicable in terms of more fundamental, smaller-scale 'causes'.

This notion complicates – perhaps it annihilates – the typical view of what an experiment in science consists of. In the double-slit experiment, say, our instinct is to regard the *phenomena* as the motions of electrons and photons along particular trajectories, or wave-like interference between them, or something of that sort, and to regard the observations – the particle-like impacts on a screen, perhaps arranged in interference patterns, perhaps not

– as the outcomes of those phenomena. Bohr asserted that instead *the entire experiment is the phenomenon that we must understand*. Whether we have one slit open or both of them, or whether we have a particle detector lurking in one slit or not, are not experiments that explore different manifestations of the same underlying phenomena. They are *different phenomena*. No wonder we get seemingly contradictory outcomes, because we are looking at different things. We should no more expect to find the same behaviour when we put a flame to a sheet of paper and to gold foil.

This intellectual strategy is both breathtaking and evasive: a bold shifting of the goalposts. From one direction it looks like cheating. In one experiment we get one outcome, but with an apparently minor modification of the apparatus we get another. And yet here is Bohr saying it's no good asking why that little change created such a different result, because we're not looking at the same thing at all in the two cases. Because the outcomes are different, we declare that the processes themselves are fundamentally different – even though the constituents of the two experiments appear to differ almost trivially (we've placed the detector *here*, not *there*). But Bohr's distinction helps to focus us on the right question. What has changed crucially, he says, is *the way we look*. So instead of trying to figure out what has made the difference to the outcome in terms of 'where the particle went', we should be asking 'Why does it matter how we look?'

This in turn prompts a deeper question: 'What information have we gained in this case that we did not have in that case?' I believe that it is *this* question, not 'Which path is the particle taking?', that may eventually lead us to a better understanding of quantum mechanics.

Bohr's prescription is extremely severe. In fact, it is more or less impossible to respect, even if you believe it. Scientists talk about electrons all the time as though they were little balls: jumping between atoms and molecules, coursing down metal wires, leaping across the void. It would be nice to be able to say that these are just convenient fictions, like the idea of the atom itself as a mini solar system of nucleus plus orbiting electrons, and that we know things aren't really that way at all. But the electron-as-cricket-ball is more than a convenient fiction. It works too well to be just that. In some situations, there seems to be no violence done to the science to think of electrons in this manner. This is one of the most challenging, if not infuriating, aspects of quantum theory: it seems to demand that we regularly disregard its insistence on what can and cannot be said. Our experience of the world – of electrons, as well as cricket balls – encourages us to ignore such mental hygiene, and to demand the right to draw pictures.

Every experimenter investigating the quantum properties of photons, for example, has to imagine particle trajectories as being real, objective phenomena when they design an experiment. They will assume that photons will follow straight-line paths as they travel through space, and figure out where to put their mirrors and lenses accordingly. At the end of these paths they will typically place a detector. The Bohrian purist would say 'You have no right to talk or think about the paths of the photons before they reach the detector. Until they are detected, the path has no meaning.' To which the experimentalist might respond 'Who cares? If I do the experiment this way, it works!' As Roland Omnès says, quantum physics involves 'an experimental physics whose modes of reasoning superbly ignore the interdictions pronounced by the theory it is supposed to check'. Bohr, he says, 'forbids too many things to the experimentalist for him to do his job'.

The experimenter might want to go further: to say to the Bohrian, 'You doubt that the photon goes in a straight line? But if I put a detector here, right in its path, what do I measure? A photon! If I move it an inch further down the path, I still detect a photon. And so on, all the way to the detector. But if I move it an inch to the side, there's nothing. Doesn't that satisfy your definition of a trajectory?'

You can perhaps guess now what the Bohrian will say in response: 'That proves nothing, because it is not the same experiment as the one with the detector at the end. It is a different phenomenon.'

It seems like an impasse: you can't prove the assumptions made in setting up the experiment by doing that experiment itself, but only by doing a different one.

Does this sound slippery? You're damned right it is.

Omnès offers a neat way out. He says, sure, one can never assert it is *true* that, in the experiment in question, the photon takes straight-line paths (if that's the design) until it gets to the detector. However, it is possible to show, using the principles of quantum mechanics, that if you assume that this is actually so, there is a vanishingly small chance that you will encounter any logical inconsistency. And this, Omnès argues, is one of the minimal requirements of an interpretation of quantum mechanics: not that you can prove it to be 'true' (whatever that means) but that you can show it to be consistent.

•

Perhaps, though, we're still not being clever enough with our experiment. Nature seems to 'know' if we're making a measurement or not – of the path of a photon or electron through the double slits, say – and changes its behaviour accordingly. It's as if nature can sense whether or not we are trying to spy on the photon's path.

Well then, let's outwit it!

Here's what we'll do. We will trick nature into showing its hand by waiting for it to make a choice – one slit or both slits – *before* we conduct a path measurement.

That's to say, we won't try to detect the photon's path until after it has passed through the slits. It's not sufficient simply to set up a detector far behind the slits, for nature seems somehow to know in advance whether it's there or not. No – we won't actually *put* it there until we know for sure that the photon has passed through the slits. Surely nature doesn't have some magical window into our intentions?

This isn't an easy experiment to do, because a photon is travelling at the speed of light. There's not much time, between its passing through the slits and its hitting the screen, in which we might whip out a detector and discern its path. But with modern optical technologies such ultrafast sleight of hand becomes possible. This is known as a delayed-choice experiment.

Einstein was the first to propose something of this sort: a thought experiment in which we leave the crucial issue of *how* we make a measurement to the last instant, after we might expect the outcome to have been already committed. How then would Bohr's 'observer-determined reality' fare?

Bohr confidently asserted that this would do no good. Nature would not be fooled. It makes no difference whether we specify the experimental arrangement in advance or postpone it until the particles are already 'in flight' – until, according to the classical view, they have already decided which path to take. Still we'll see the same result as we do with a conventional experiment.

He felt able to assert this because it seemed to be what quantum mechanics predicted. But it made no sense! As John Wheeler later pointed out, it seems to imply backwards causation: an event happening at one time having an influence on an event at an earlier time. By finding out whether a photon, having already passed the slits, went through one or both, we appear to be determining which of those was the case. As Wheeler put it, by moving such a delayed photon detector in or out of the apparatus, we 'have an unavoidable effect on what we have a right to say about the *already* past history of that photon'.

Notice how carefully Wheeler expresses this: that we have an effect not on the past history of the photon, but on what we have *a right to say* about it. Because, as he went on to explain, we don't *really* alter the past history; rather, we have to alter our entire view of what the phenomenon is that we're observing:

> In actuality it is wrong to talk of the 'route' of the photon. For a proper way of speaking ... it makes no sense to talk of the phenomenon until it has been brought to a close by an irreversible act of amplification [that is, a measurement on a classical instrument]: 'No elementary phenomenon is a phenomenon until it is a registered (observed) phenomenon.'

If, as Bohr said, the quantum experiment is not probing the phenomenon but *is* the phenomenon, we can't speak about that phenomenon having taken place until the experiment is done and the measurement is made – until the needle on the meter registers a reading. In order to be able to speak of it as an actual event, we have to *see* it.

We are used to the idea of machinations happening beyond our perception. Our cells are busy doing their biochemical business, making proteins and fighting infections and so on. Molecules in the air are colliding invisibly; as they impinge unseen on a surface in their countless billions, they impart a tangible pressure. We can intervene in these phenomena and make measurements, but we are justified in assuming that the microscopic phenomena continue regardless of whether we do or not.

Yet in the view of Bohr and Wheeler, there are no fundamental quantum phenomena *about which we have any right to speak* until we measure them. To the question 'What was happening to the photon between its emission from the laser and its detection?', we can't simply reply 'I don't know, I wasn't looking.' We have to say '*Because* I wasn't looking, that question has no meaning.' Or perhaps, 'Well, once I've made the measurement, then we can talk about it' – rather as the result of a particular football match becomes a valid concept only when it's all over.*

This picture is striking because it doesn't depend on the physical act of making a measurement. There's something deeper at work, relating to our *gaining knowledge*. Carl von Weizsäcker, whose perceptiveness about quantum theory was perhaps second only to Bohr's, put it astutely (the italics are mine):

> It is not at all the act of physical interaction between object and measuring device that defines which

* Bohr would, I like to think, have approved the analogy; he was a goalkeeper, and his brother Harald, a mathematician, played in the Danish national team.

> quantity [for example, which path] is determined and which is undetermined, but *the act of noticing.*

•

Bohr was characteristically confident in his prediction about the delayed-choice experiment. But was he right? The only way to know was to do the experiment. That became possible once Wheeler, in the late 1970s, proposed an experiment with laser photons that mimicked the double-slit experiment while avoiding the awkward business of how on earth to stick a detector into the arrangement that could detect with certainty which path a photon took *after* it had apparently taken it.

Here's Wheeler's scheme. We shine a laser beam of photons at a mirror inclined at 45° to its path. The mirror (M_1) is 'half-silvered' – partly reflective and partly transparent, so that it reflects 50% of the incident photons (at random) and allows the others to pass through. Thus it acts as a beamsplitter, dividing the beam into one (A) that continues in a straight line and another (B) that bounces off at a right angle. We position mirrors to reflect the two paths back towards a crossing point. We monitor which route the photons take using two sensitive photon detectors D_A and D_B. They confirm that half of the photons pass along A, and half along B, at random.

Now we place another half-silvered mirror (M_2) at the crossing point of the two beams. This induces interference between the two beams, creating a light-and-dark pattern of interference fringes – and we lose the ability to say if the photons went along path A or path B. We can arrange for the two detectors, while still positioned in the lines of paths A and B, to be just at the points where the interference pattern has a bright and a dark band respectively. Then when photons pass singly through the apparatus, D_A will register the arrival of a photon 100% of the time, while D_B will register none.

So the statistics of photon detections are completely different, in a totally predictable way, depending on whether mirror M_2 is in place (D_A = 100%; D_B = 0) or not (D_A = D_B = 50%). In the latter case we know for certain which path each single photon took, since it will show up either at D_A or D_B, with equal probability. But with M_2 in place we can't assign it a single path: D_A will register the photon's arrival with 100% probability come what may, implying that it took both paths and interfered with itself.

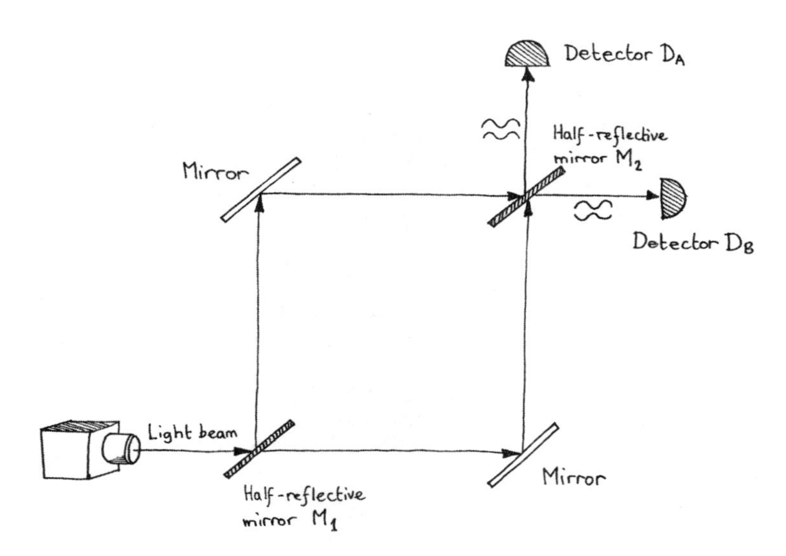

John Wheeler's delayed-choice experiment.

To make this a delayed-choice experiment, we must be able to insert M_2 *after* we can be sure that a photon has passed M_1, when it should have committed itself exclusively to one or other route – but not both, because M_2 wasn't in place at that stage. Using modern fibre-optic technologies, it's possible to do this. If Bohr was right, we'd still then see the detection statistics corresponding to interference (D_A = 100%; D_B = 0), even though the apparatus had the 'interference-free' arrangement when the photon encountered M_1 (the event equivalent to passing through the slits in the double-slit experiment).

The first experimental implementations of Wheeler's arrangement were made in the late 1980s. Many variations have been tried since. They all show that indeed it makes no difference when we intervene, so long as we do so before a measurement is made. Nature always seems to 'know' our intentions. Or to put it again in John Wheeler's less spooky but no less perplexing way:

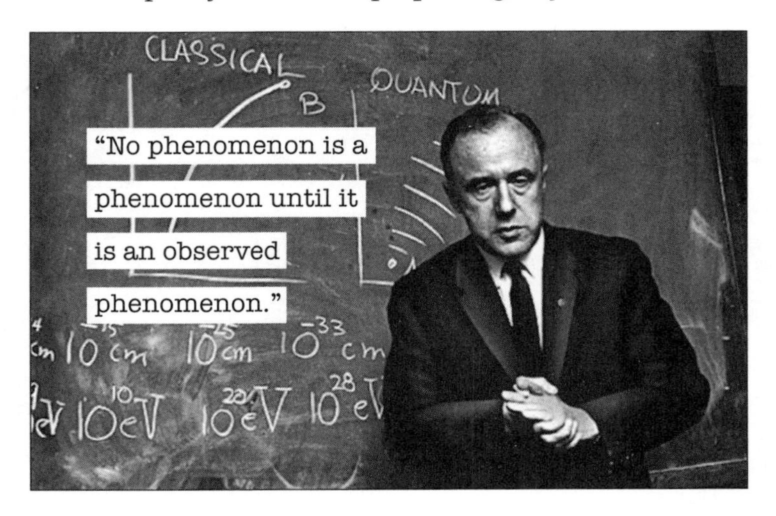

•

How is this possible? What is *really* going on when we make a measurement? The Copenhagen Interpretation commands us not to ask such things. But we can now express more precisely what it is that this prohibition sweeps under the carpet.

Forget about 'reality' for the moment – it's too tricky a concept (which is no news to philosophers). Let's just ask what happens *in theory* when a measurement is made. Before that's done, a quantum system behaves in accordance with the Schrödinger equation, which describes how the system's wavefunction changes as time passes. Crudely speaking, the theory simply says that this change is smooth and wave-like. At one moment the wave amplitude is big here and small there, at the next moment the reverse is true.

One of the properties of a quantum system is that this change over time preserves distinctions between states. What I mean here is something like this. Classically, if two states start off being different, and both experience the same influences, they stay different. Say I throw two identical tennis balls up into the air at the same angle but at different speeds. The slower one will always fall to the ground sooner and nearer than the other, and at a time and place that is wholly predictable. This seems obvious – in essence, it's saying that systems don't change their state for 'no reason'.

This principle is not quite the same for quantum systems, because they are governed by probability and are prone to randomness. We can only calculate the chance of a 'quantum tennis ball' landing in different locations at different times; we can't be sure of the actual outcome in a given experiment. But what we can say is that the sum of all the probabilities of possible outcomes of a quantum event must be equal to 1. That's simply saying that, of all the things that *could* happen, one of them *must* happen.

This is really a statement about the *information* in the system: it is never lost. Here's how such loss can happen in situations governed by probabilities of outcomes. Say you have two coins concealed under two cups, and you know that they are either both heads or both tails, both with 50% probability. Now someone, in a move unseen by but announced to you, flips one of the coins. You now know, with the same equal probability, that either the lefthand cup conceals a heads and the righthand one a tails or vice versa. It's possible to show that, in mathematical terms, no information has been lost.

But what if instead someone shakes the coins in the cups to randomize their orientation? Now you have a 25% probability of each configuration: head/head, tail/tail, head/tail, tail/head. And again, you can show that mathematically you've lost information. (Crudely: in the previous cases you could confidently exclude some configurations, but now you can't.)

A process that conserves information in this way is said to be unitary; shaking the cups, in contrast, is a non-unitary transformation. And the way that quantum systems evolve through time according to the Schrödinger equation is strictly unitary. But quantum measurement seems to violate unitarity: it imposes a violent rupture on the smooth evolution of the quantum wavefunction.

Before measurement, then, the system is fully described by a wavefunction from which one can calculate the various probabilities of the different possible measurement outcomes. Let's say that the system is in a superposition of possible states A, B and C. Then, according to quantum mechanics, the wavefunction can do nothing except continue evolving in its unitary way, preserving these three possible states.

But measurement does something else. It 'collapses' (Heisenberg's original word was 'reduces') those possibilities, expressed in the wavefunction, to just one. Suppose that, before the measurement, the probabilities of finding some property of a quantum object with the values corresponding to states A, B and C are 10%, 70% and 20% respectively. When we make a single measurement on the object, we might find that we get the result C. What happened to A and B? We are forced to assert that the probabilities have now changed: that for C it is 100%, and for A and B it is zero. What's more, we can't get A and B back: if we repeat the measurement, we'll keep getting C.*

What causes this abrupt change? It's not something predicted by the theory. There is nothing in the Schrödinger equation that allows or accounts for wavefunction collapse. You can't start off with states A, B and C and end up with just C by evolving the wavefunction in a unitary way. To put it somewhat crudely but I think aptly enough, if you have a mixture of red, yellow and blue paint, you can't somehow adjust the blend to end up with pure blue. And if some magical operation *does* leave you with pure blue, there's no way (without going back to make a fresh mixture) you can ever regain a tinge of redness or yellowness.

* The only way to recover A and B is to prepare the quantum object again in the same original state. It won't just happen of its own accord. When the object is newly prepared this way, A, B and C again become possible outcomes of a measurement.

Here, then, is the problem. The fundamental mathematical machinery of quantum mechanics is unitary: the Schrödinger equation which describes how a wavefunction evolves through time prescribes that this evolution is only and always unitary. Yet every experiment ever performed on a quantum system which sets out to directly measure some property of the system induces what we are forced to call 'collapse of the wavefunction': it gives a unique answer. And this is necessarily a non-unitary process, and therefore inconsistent with what wavefunctions seem able, in theory, to do.

So we have every reason to suppose that quantum mechanics is unitary, and yet we observe non-unitary outcomes of experiments. This is why the measurement problem is so upsetting.

•

The early Copenhagenists insisted that the apparently non-unitary collapse of the wavefunction is simply what measurement is all about: they tried to neutralize the problem by making it a kind of axiom. But this was not much better than saying that wavefunction collapse happens by magic. There was no theory for it.

To Bohr, wavefunction collapse was virtually *emblematic* of the distinction between the unitary quantum world and the everyday reality in which we make observations and measurements. Measurements must be classical by definition: they require some big apparatus with which humans can interact. From our perspective the world is made up of phenomena – things happen – and a phenomenon only exists when it has been measured. Wavefunction collapse is simply a name we give to the process by which we turn quantum states into observed phenomena.

Wavefunction collapse is then a *generator of knowledge*: it is not so much a process that *gives us the answers*, but is the process by which *answers are created*. The outcome of that process can't, in general, be predicted with certainty, but quantum mechanics gives us a method for calculating the probabilities of particular outcomes. That's all we can ask for.

•

Without measurement there seemed to be nothing to destroy the unitary evolution of the Schrödinger equation, with all its superpositions and multiple possibilities. So what happens at the macro scale when we don't look? Einstein once expressed his exasperation at the implications of Bohr's position to the young physicist Abraham Pais. 'I recall', Pais later wrote, 'that during one walk Einstein suddenly stopped, turned to me and asked whether I really believed that the moon exists only when I look at it.'

This focus on 'looking' as the source of wavefunction collapse carried with it an implication that it was not exactly 'measurement' – the interaction with a macroscopic instrument – that mattered, but as von Weizsäcker said, 'the act of noticing'. It seemed to demand a consciousness to register the event.

Or did it? Should we consider the wavefunction to have collapsed within the measuring device, or within the brain of the human experimenter? At what point in the chain from quantum event to macroscopic measuring device to observer reading the result and noting it in a lab book do we consider collapse to have occurred? Werner Heisenberg pondered this problem, and the point at which we separate the quantum from the classical world became known as the 'Heisenberg cut'. Where is it?

Bohr and Heisenberg disagreed about that. For Heisenberg the 'cut' was not a physical boundary at which something (like wavefunction collapse) happens, but a place where we rather arbitrarily choose to divide the measuring system from what is being measured. We're pretty free to decide where the cut goes, he said, so long as it is not *too* close to the quantum object, and provided that we construct the mathematical description of the process accordingly.

Bohr wasn't happy with that shiftiness. He felt that the location of the cut depends on which questions we choose to ask in an experiment, but that once we've selected those questions then the cut is fixed. It corresponds to the point in the process where we can obtain clear answers to those questions: a place where we might say 'wavefunction collapse happens here'. This attests to Bohr's conviction that in the end it all comes down to the matter of what your experiment looks like. Until you specify that, you can't really talk about anything much.

•

There is something infuriatingly unreachable about Bohr's perspective. If one were to have asked him 'So, does quantum mechanics break down when you make a measurement?', it's far from clear (to me, anyway) that he would have said 'Yes.' But wavefunction collapse is non-unitary, you might protest, and so it conflicts with the Schrödinger equation! Ah, Bohr might reply, but wavefunction collapse is just a concept we invoke when we make a measurement.

And measurement is necessarily a classical process, so we can't go applying quantum math to it. Measurement is precisely how we acquire knowledge – so if measurement wasn't classical, we wouldn't be able to get any knowledge about a quantum system by experiment.

You might then be inclined to exclaim 'Damn it, Bohr, you're avoiding the issue!' – and go stomping off to look for another way to think about quantum mechanics. It's completely understandable that many people did so, and are still doing so.

There are many ways of
theory (and none of them

interpreting quantum
quite makes sense)

The Copenhagen Interpretation is sometimes said to be the 'orthodox' vision of quantum mechanics. This isn't really true. It might be the most popular interpretation, but not overwhelmingly. There is no quantum orthodoxy.

Neither, as I've said, is there even a unique, consensus version of the Copenhagen Interpretation. I have been rather cavalier with the phrase, and will continue to be, for this book is too short for perpetual qualifiers. Some deny that there was ever any shared core belief among 'Copenhagenists', and maintain (it seems to me a plausible account) that the entire notion of a 'Copenhagen interpretation' was invented largely in the 1950s by Werner Heisenberg, perhaps to re-insinuate himself into the 'Copenhagen family' after his fateful, bitter meeting with his mentor Bohr in the occupied Danish capital in 1941 to discuss the German atomic bomb project. But if you're looking for a view from Copenhagen, you're best advised to get it from Niels Bohr himself. And if you want to argue with it, you must argue with Bohr.

That's not much fun, as Einstein discovered. Bohr's writings are ponderous and often hard to fathom. He wasn't naturally gifted as a writer – he would draft and redraft endlessly without much obvious benefit to the prose. But the challenge in reading Bohr comes also from the fact that he took such tremendous care to say what he meant. As he said:

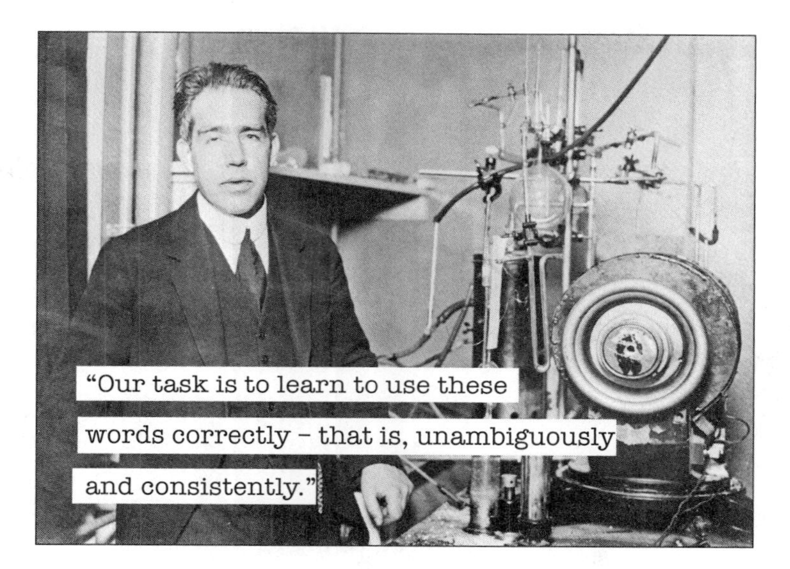

The problem is that in quantum mechanics it is almost impossible to be unambiguous and consistent, to say what you mean, or perhaps even to know what you mean, because you are dealing with concepts that defy language. Here is how von Weizsäcker tactfully put it:

> Bohr's writings are characterized by a highly implicit and carefully balanced mode of saying things, which makes reading his work rather arduous, but which is in harmony with the very subtle content of the quantum theory.

Bohr could be stubborn and dogmatic as well as rather cryptic. But he deserves praise for defining the limits of what can be said with confidence. Some suspect that his intuition surpassed anything he could justify mathematically, or indeed semantically. As von Weizsäcker said, 'Bohr was essentially right but he did not know why.'

It's clear now that Bohr's view of quantum mechanics cannot be 'right' in any absolute sense – it is too restricted, and he could not have known some things that we now understand. But about where the problems lie, he was right. And probably, yes, he did not know why.

Still, some researchers regard the rather equivocal predominance of the Copenhagen Interpretation as mere historical contingency at best, and as the result of effective (even aggressive) marketing at worst. The Nobel laureate physicist Murray Gell-Mann accused Bohr of having 'brainwashed' a generation of physicists into thinking that the problems of quantum mechanics had been solved: the Copenhagen Interpretation has, he said, a 'tranquillizing' effect that induces an uncritical stupor.

Even if you don't object to the Copenhagen Interpretation, you have to wonder whether its hegemony really is due to anything other than chance – or to the canny machinations of its advocates, marshalled by the indefatigable Bohr. The physicist and philosopher James Cushing has argued that one could equally imagine at least one of the rival interpretations having been formulated instead in the 1920s – and then, having won support from the likes of Einstein and Schrödinger (who never accepted the Copenhagen view), becoming the standard story instead. But that's not how it happened. 'The Copenhagen Interpretation got to the top of the hill first', says Cushing, 'and to most practicing scientists there seems to be no point in dislodging it.'

I have so far repeatedly drawn on the Copenhagen Interpretation to explain how we should *really* understand the alleged weirdness of quantum mechanics. I've done this not because (I don't *think* it is because) I have been brainwashed into a conviction that this interpretation is correct, nor even because I harbour a suspicion that it is. It is because Bohr's picture offers the clearest way to see where the interpretational problems lie, and to distinguish between what we can say for sure and where such confidence has to be relinquished. It has the virtue of being explicit about the limits of our knowledge. We know that measurements of a quantum system *seem* to collapse the wavefunction. We most certainly *don't* know how, or why, or indeed *if* that actually happens.

This virtue is also the weakness of the Copenhagen Interpretation. It prohibits further probing, and so leaves wavefunction collapse a mystery – and moreover, one to which we can admit no solution even in principle. The Copenhagen view is consistent, certainly – but it's not hard to be consistent if you refuse to entertain awkward questions. It is entirely understandable that some see Bohr's doctrine as a counsel of despair, or alternatively as a cheap cop-out. It demands that, at the moment of measurement, we accept that the universe does something not really distinct from magic.

So what are the alternatives? Cushing's point about the Copenhagen Interpretation is more than a mere acknowledgement of contingency. It implies that there is *no obvious way* to deal with what strikes us as strange about quantum mechanics. Nothing we try will make it go away. It's for this reason that the proliferation of interpretations is not a failing of quantum mechanics, but a necessity. We need different perspectives, just as we need to look at a sculpture from many different angles to appreciate it fully.

Some say that it can be only a matter of taste what we choose to find unacceptable in any of these interpretations. That is probably true right now. At any rate, I suspect that the reasons why researchers align themselves with one school of thought or another are far more nebulous and subjective than they might care to acknowledge. They might say that a particular interpretation 'makes sense' to them, and could probably adduce logical-sounding reasons for that. But there's surely a lot of gut feeling involved. In the end, the perspective that we find persuasive or satisfying may be the one that best flatters our preconceptions and prejudices. In what they thought about quantum theory, we catch a glimpse of the personalities of Einstein, Bohr, Heisenberg, Schrödinger, Wheeler, Feynman. In what we feel about it, we doubtless reveal a little of ourselves.

•

The impulse to restore an objective reality beneath the opaque, symbol-strewn carapace of quantum mechanics has always been deeply felt. One of the most inventive ways of doing this was proposed by the American physicist David Bohm, who worked under J. Robert Oppenheimer in California in the 1940s before collaborating with Einstein at Princeton. Bohm turned Louis de Broglie's 'wave-particle duality' of quantum objects from a mutually exclusive choice into a supportive partnership. He supposed that a description of the mechanics of quantum entities needs both a particle *and* a wave to be literally present. The particle is as definite an object as any classical one, and the wave guides its motion – it is sometimes called a 'pilot wave'. The particle's movement is then fully deterministic – we can consider it to have a definite position and trajectory – but it incurs some unknown, random variation due to the properties of the wave. So any uncertainty about the particle's properties is of the classical variety: we just don't (and can't) know all the fine details.

This pilot wave is a rather mysterious thing: a vibration in some pervasive and exquisitely sensitive (and, I must stress, purely hypothetical) field called the quantum potential. It is able to guide the particle without exerting anything we'd conventionally recognize as a force, which means that it doesn't require any source of energy. Neither does its influence decline with increasing distance, as do ordinary forces like electromagnetism and gravity. What's more, the wave, being a spread-out affair, can collect information about its environment and 'feed' that instantaneously* to the particle to direct its motion accordingly.

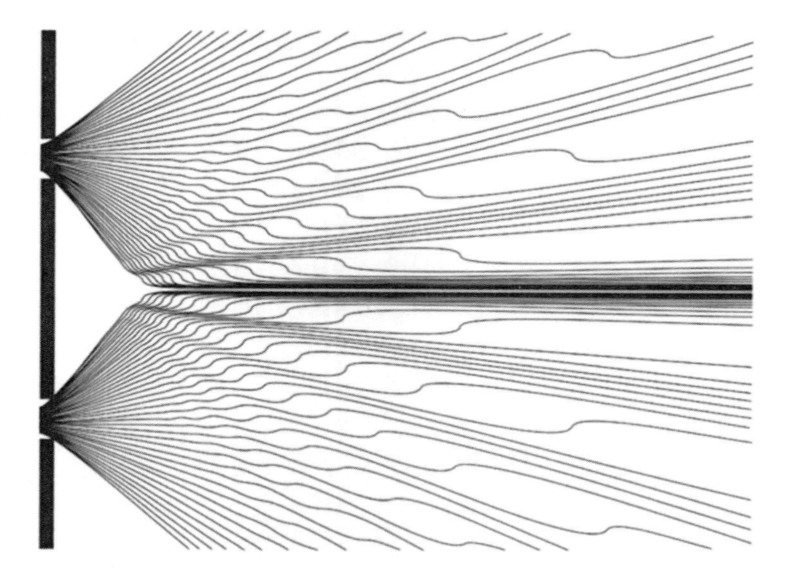

The trajectories of particles in a double-slit experiment under the guidance of Bohm's 'pilot wave'. These mimic a wave-like interference pattern.

* This action of the quantum potential does not, however, violate the injunction of special relativity that no signal can be sent faster than light. The quantum potential is so sensitive that any attempt to manipulate it to send a message has a completely unpredictable effect, garbling the content.

This motion needn't resemble the kind of smooth, straight-line trajectories common in classical mechanics, and it enables the kind of experiment-dependent and apparently non-localized behaviour of the quantum particle. If we try to probe the path of a particle, we disturb the quantum potential in a way that destroys the interference-like behaviour arising from the pilot wave, accounting for what is seen in a quantum double-slit experiment.

Thus Bohm brought a classical picture of particles back into the quantum microworld. But the cost of restoring this underlying reality is to package all the 'quantumness' into the almost miraculous quantum potential.

There's nothing obviously impossible about the idea of a quantum potential. But neither is there a shred of evidence for it. And for Bohm it had tremendous powers beyond what quantum mechanics seems strictly to require. He felt that the 'active information' that it could transmit to a particle had parallels with the activity of the mind, turning the entire universe into something resembling a conscious organism. This confers a unity that Bohm called the 'implicate order', which underpins the 'explicate order' accessible to the senses. Thought exists in the cosmos as a holistic entity akin to the quantum potential, which it would, he said, be 'wrong and misleading to break ... up into my thought, your thought'.

This quasi-mystical view of reality has made Bohm popular with the New Age movement – sometimes to the detriment of what were rather profound, if somewhat abstruse, reflections on the message of quantum theory. His influence and legacy are important, and the de Broglie–Bohm interpretation (as it is sometimes called) has advocates today. But it is hard to see where the gain lies.

It looks reassuring that the model restores an underlying reality of particles existing in particular places – but this happens only at the cost of giving them quantumness by fiat via the quantum potential. Only a minority of physicists and philosophers consider that a good bargain. Even Einstein, who was certainly keen to win back objective reality from quantum theory's apparent denial of it, found Bohm's idea 'too cheap'. One objection is that the particle trajectories it predicts are bizarre and at odds with any ever observed. Others say that the paths only *look* that way, because the non-locality of the quantum potential gives us an unreliable view: an exculpation some might consider a little too convenient. At the very least, however, Bohm's description gives us an indication of the kind of magic we would need to recover classical-like particles.

•

The problem with the collapse of the wavefunction is that, as we saw, quantum mechanics contains no prescription for it – it has to be added by hand. Well then, might something be missing from quantum mechanics? If collapse is what we seem to 'see', why not add in some math to describe it? Isn't that, after all, what we normally do in science?

Oh, if only it were so simple! Every measurement ever made on a quantum system is consistent with the Schrödinger equation *as it stands*, without any additional tinkering. If we stick another piece onto the equation that forces the wavefunction to collapse, won't we screw up this lovely concordance?

Not necessarily. In 1985, the Italian physicists Giancarlo Ghirardi, Alberto Rimini and Tullio Weber (denoted GRW) proposed a modification to the Schrödinger equation that, with a judicious choice of the mathematical parameters, could allow it to retain its validity for the microscopic world while compelling wavefunction collapse in the macroscopic world.

GRW added to the equation a term describing a random process that, over the course of time, repeatedly 'prods' a quantum superposition until it jumps suddenly into a single state with a fixed, precise location. It's a bodge, really: the researchers just figured out what kind of mathematical function was needed to do this job, and grafted it on.

The point is that this 'localization term' can be freely tuned to adjust the timescale on which collapse takes place. It ensures that the bigger the object, the faster localization happens. With an appropriate choice for the 'strength' of their added effect, macroscopic systems can be localized virtually instantaneously whereas for typical quantum systems like an electron the collapse won't happen spontaneously for billions of years, meaning that in practice we'd never expect to see it. Collapse of the wavefunction for a microscopic particle *does* happen, however, once it is coupled to a macroscopic apparatus in order to make a measurement.

If you think this seems like special pleading, you'd be right. But that's no objection. There's no obvious reason why the 'collapse' term in the revamped GRW Schrödinger equation shouldn't just happen to be tuned to give us microscopic quantumness on the one hand and macroscopic classicality on the other.

What's more of a problem is that there is absolutely no evidence that such an effect exists. You might say 'But wavefunction collapse is precisely what we *do* see!' Yet it isn't. We see the unadulterated Schrödinger equation working just fine for quantum systems, and classical, deterministic physics working for big systems. Wavefunction collapse is simply a conceptual fudge for cobbling the two together: it's not an observed, physical process like, say, radioactive decay of an atom.

GRW's modification of quantum mechanics implies, however, that wavefunction collapse *is* like such a process. If so, it is a process hitherto unknown in physics, and one that we should expect to be able to detect. The GRW model is now just one of a general class, called physical collapse models, that assume something of this nature.

Another physical collapse model was devised in the 1980s and 90s by the British mathematical physicist Roger Penrose, and independently by the Hungarian physicist Lajos Diósi. They suggested that collapse might be induced by the disrupting influence of gravity. In this view, classical behaviour is directly a result of size – more precisely, of mass. Loosely speaking, the idea is that if objects are big enough to exert an appreciable gravitational force, then one object 'feeling' the position of the other via gravity amounts to a measurement-like influence that will destroy quantum superpositions of states.

Physical collapse necessarily means that the unitary nature of quantum mechanics – loosely, the idea that states that are initially distinct always remain distinct – breaks down. For Penrose, there's nothing sacred about quantum unitarity: it doesn't have to apply 'all the way up'. In fact, he says, it clearly doesn't, since cricket balls can't be placed in quantum superpositions. He argues, then, that physical collapse isn't a clumsy fix to avoid interpretational difficulties, but is, just like any other scientific hypothesis, motivated by what we observe.

This idea has the considerable virtue that it is amenable to experimental testing – it is not so much an 'interpretation' of quantum mechanics as a straightforward extension of the theory. Some researchers are hoping to test the Penrose–Diósi model by looking for quantum effects in objects big enough to be susceptible to gravitational influences. These plans are often ambitious, demanding extreme environments and incredibly sensitive measurements. Markus Aspelmeyer of the University of Vienna and his colleagues hope to conduct an experiment called MAQRO on a space satellite in zero gravity, where they would place a particle about 100 billionths of a metre in size – that's big in quantum terms – in a quantum superposition and then use lasers to probe how quickly the superposition disappears, compared to the same situation in Earth's gravity. In the Penrose–Diósi model this should happen at a different rate than in regular quantum mechanics.

•

The most controversial – one could fairly say notorious – way of dealing with wavefunction collapse is to do away with it altogether: to treat it as an illusion that only *seems* to select one option from many at the macroscopic scale. I deal with this so-called Many Worlds or Everett Interpretation of quantum mechanics later, but its key attribute is that it refuses to recognize any limitations at all to the applicability of quantum mechanics. The theory applies as much to the entire universe as it does to individual photons and electrons, so that the universe too can be assigned a wavefunction.

It's not entirely clear what it can mean to posit an expression like that, since it could never be written down even in principle. (As we'll see, the problems with the Many Worlds Interpretation go far deeper.) But nevertheless the notion of a universal wavefunction is popular with cosmologists, for the perfectly valid reason that in the earliest moments of the Big Bang the entire universe was smaller than an atom and surely needs to be considered, in that moment, a quantum-mechanical entity.

Such a wavefunction would need to contain within it every state of the universe conceivably possible. Yet not all of these are realized – specifically, at large scales only certain classical states are allowed. Why is that? In the so-called Consistent Histories Interpretation of quantum mechanics, developed in the 1980s by the physicist Robert Griffiths and subsequently and independently by Roland Omnès and by Murray Gell-Mann and James Hartle, we can narrow down the options using sheer logic. That's to say, even though quantum mechanics forgoes any choice of outcomes – *this* or *that* – in favour of just their probabilities, we can still reasonably say that anything that happens has to be consistent with what came before. Such a criterion of logical consistency means that *not every history can be assigned a probability.* Quantum mechanics allows us to distinguish between the two possibilities: either a particular history is consistent or it is not.

This view allows us to sharpen what we can say about the double-slit experiment. Typically, it's said that if we don't measure the particle's trajectory (and we therefore see an interference pattern emerge from the particle impacts on the screen), we can't say which slit the particle passed through, and have to allow that it must have gone through both at once. But the Consistent Histories view is that in this case we can't meaningfully talk about such trajectories at all, since it is formally impossible, using the math of quantum mechanics, to assign any probability to the particle's taking either route. It's not that 'because the particle went through both slits, there's an interference pattern', but rather, 'the outcome in which there's an interference pattern contains no meaningful definition of particle trajectories'. Trying to impose some fuzzy microscopic interpretation of the observed results is simply the wrong way around; for certain outcomes, no microscopic interpretation is *logically* meaningful.

The Consistent Histories Interpretation offers a clear way to think about what Bohr made speakable and unspeakable in quantum mechanics. We don't ban some questions simply because we don't know what to say about them, but instead recognize that quantum mechanics has no math that can provide an answer: it's rather like expecting simple arithmetic to tell us what an apple tastes like. In that much, Consistent Histories offers a valuable tool. But it stops short of supplying a physical picture that improves on other interpretations – which is why it is not exactly inconsistent with some of them.

•

The Hungarian mathematical physicist John von Neumann was one of the first to make wavefunction collapse an 'official' component of quantum mechanics, incorporating it into his 1932 textbook on the subject. He pointed out that the collapse happens through the intervention of an observer, and so figured that it must have something to do with the act of observation itself. This led his compatriot Eugene Wigner to hypothesize that collapse stems from *conscious intervention* in the quantum system – that it is produced by our own minds. It was, to be sure, something of a desperate idea: an attempt to curtail what might otherwise become an indefinite postponement of the time and place where quantum becomes classical.

Wigner illustrated the idea with a thought experiment now known as Wigner's Friend. Suppose that Wigner conducts a measurement on a superposition of quantum states, the possible outcomes of which are such that an observable flash is produced (indicating that the quantum system is in a state with one particular wavefunction) or not (indicating that it is described by a different wavefunction). It is only when the flash is registered (or not) that one can meaningfully decide which outcome has been realized, and can thus consider the superposition of states to have collapsed.

Now suppose that this experiment is conducted after Wigner leaves the lab, leaving his friend to make the observation. If we consider the situation from the quantum-mechanical point of view, Wigner can't meaningfully say that the wavefunction has collapsed until his friend tells him the result. It's not just that Wigner doesn't know the outcome until that point; quantum theory offers no prescription by which Wigner can speak about those alternatives as real events at all.

In this view, Wigner's friend is herself in a superposition until Wigner collapses it by extracting the information. But now we seem locked into an infinite regress. Is Wigner himself in a superposition of states until he tells the result to his other friends, who are anxiously awaiting the news in the next building? Does collapse spread over the planet with the news of the result? Which observer 'decides' when wavefunction collapse occurs?

There are all kinds of other problems with this idea. What, for example, constitutes a conscious observation? Does a dog seeing a meter reading from a quantum experiment – or perhaps just observing a light bulb switching on, which even a dog can register and in some sense report – bring about collapse of the wavefunction? Indeed, even fruit flies can be trained to respond to the kinds of stimuli that could signal the outcome of a quantum experiment ...

At what point, then, does consciousness enter the picture? And how, in any case, can we reasonably make the mind responsible for reducing all the quantum probabilities to a single certainty while we still lack a theory of mind, brain and consciousness?

In particular, mind-induced collapse seems to demand that we attribute to the mind some feature distinct from the rest of reality: to make mind a non-physical entity that does not obey the Schrödinger equation. How else could it do something to quantum processes that nothing else can?

Perhaps most problematically of all, if wavefunction collapse depends on the intervention of a conscious being, what happened before intelligent life evolved on our planet? Did it then develop in some concatenation of quantum superpositions?

John Wheeler offered an extraordinary view of cosmic evolution that depends on such consciousness-induced collapse of the wavefunction. If 'noticing' – that is, observation – doesn't just report on but actually produces phenomena, crystallizing 'what happened' out of 'what might have happened', could the presence of beings capable of 'noticing' transform a multitude of possible pasts into one concrete history? Could it be that only when we register quantum events – the interactions of countless particles in the past – do they become actual events? Wheeler offered a cosmological version of his two-path experiment (page 93) in which the gravitationally induced bending of a light path by a distant galaxy supplies photons from a yet more distant body with two possible routes to a detector on Earth: one direct, one along the bent path. A photon could have passed the 'lensing' galaxy billions of years ago – and yet, by placing a beam-splitter in front of the detector to measure if it might self-interfere or not, we determine whether we can retrospectively speak in terms of its taking one path or two.

More generally, by 'noticing' how things are today, we might be selecting which of many quantum paths they took in the past – and in this sense we become participants in the evolution of the universe since its very beginning.

It's not clear to me how this meaningfully alters anything we can say about the way the cosmos has developed. It doesn't make much sense to suppose that the moon and all the geological evidence of its presence jumped into existence the moment – who? The first *Homo*? A tyrannosaurus? – looked up and saw it. Rather, Wheeler's 'participatory universe' earns its keep as another thought experiment through which to explore what a quantum observation could mean.

•

Bohr's injunction forbidding us from speaking about any objective quantum reality beyond its effects in particular experiments only went so far, for it insisted that objectivity was restored in the classical realm. Why, though, should one set of rules switch to another? Bohr insisted that quantum and classical are fundamentally different realms of experience, and papered over the divide with the word 'complementarity'. The world, he said, consists of complementary elements that have an exclusive existence, so that we cannot know them all at once. There is a sense in which this seems to be true, but intoning power-words was not enough to excuse the discrepancies.

Another interpretation of quantum mechanics refuses this easy way out. It is called Quantum Bayesianism or QBism (cutely pronounced 'cubism'), and it was formulated by the physicists Carlton Caves, Christopher Fuchs and Ruediger Schack in the early 2000s. You could call it an interpretation that is more Copenhagenist than the Copenhagenists. Bohr had said that the purpose of quantum mechanics is not to tell us about reality but to predict outcomes of measurements. In QBism this philosophy is extended to everything there is, quantum or classical – everything, that is, outside the observer's conscious perception.

In other words, in QBism quantum mechanics is used to describe *everything* external to the observer. It's then perfectly permissible to speak of superpositions of macroscopic states or objects: Schrödinger's cat, Wigner's friend and so forth. But we never observe such things, so how can we say what they would mean? Well, in QBism you can. Here, all quantum mechanics refers to are *beliefs* about outcomes – beliefs that are individual to each observer. Those beliefs do not become realized as facts until they impinge on the consciousness of the observer – and so the facts *are specific to every observer* (although different observers can find themselves agreeing on the same facts).

This notion takes its cue from standard Bayesian probability theory, introduced in the eighteenth century by the English mathematician and clergyman Thomas Bayes. In Bayesian statistics, probabilities are not defined with reference to some objective state of affairs in the world, but instead quantify personal degrees of belief of what might happen – which we update as we acquire new information.

The QBist view, however, says something much more profound than simply that different people know different things. Rather, it asserts that there are no things that can be meaningfully spoken of beyond the self. This might sound incredibly solipsistic, but it isn't really. You could argue that it just embraces the truth of the situation we are inevitably in, locked as we are into our own consciousness. It doesn't deny the existence of anything outside of that subjective experience, but it denies us knowledge of it.

Quantum mechanics generally assumes that quantum states exist in some meaningful sense, and that the math tells us what we can know about those states. But in QBism there *are* no objective states. Rather, according to Chris Fuchs, 'quantum states represent observers' personal information, expectations and degrees of belief'. This view, he says, 'allows one to see all quantum measurement events as little "moments of creation", rather than as revealing anything pre-existent'.

If we extend this notion of the subjectivity of states beyond the world traditionally considered quantum, we find that some of the apparent paradoxes of quantum mechanics vanish. In the QBist view, Wigner's friend is in a superposition *as far as Wigner is concerned* because

Wigner hasn't yet observed her and so doesn't know what outcome of the experiment she saw. But she is not in a superposition from her own perspective, and experiences no strange 'two states at once'.

This feels like another sleight of hand. Worse, it makes the world even more intangible and unspeakable than the strictest Copenhagen Interpretation. Everything that Bohr prohibited about the quantum world – imagining some objective reality beyond what we can measure – now applies to the classical world too. Why make that sacrifice, given that an objective reality – things with properties that are fixed before we look – seems so plainly to exist at the classical scale? Isn't QBism just a retreat to unverifiable sophistry? How can it be valid to banish 'weirdness' by making *everything weird*?

But that's not the right way to look at it. For QBism is not, as sometimes supposed, taking the ultimate self-regarding step of making reality just an illusion conjured by our own minds. It is genuinely an interpretation of quantum mechanics – which is to say, it interprets the *theory* without pronouncing on what lies beyond it. It claims only that such an objective world exists, and that quantum mechanics is the framework that *we* need to make sense of it. This framework has the form it does, say QBists, because the nature of the world is such that our intervention in it matters. We affect what transpires. This doesn't mean that we determine *everything* that happens, or even most of it; indeed, we have no influence on almost anything. But where we do, we touch the nature of the reality of the world, and we play a part in what that nature produces.

QBism, then, embraces the notorious 'observer effect' in quantum mechanics in a particularly subtle way. It makes quantum mechanics the theory needed to make sense specifically of that situation in which decision-making agents like us interact with some tiny fragment of the universe that captures our attention.

You might complain that this navigates the conundrums and paradoxes of quantum mechanics by diminishing the theory: QBism declines to talk about reality beyond our experience of it. But how, virtually by definition, can we ever hope to know more than that? And QBism is not utterly silent about what lies beyond: about what makes the world the kind of place where quantum mechanics is required. In Fuchs' words, the glimpses that we are permitted suggest that, rather than being fixed into some rigid, deterministic mechanism, there is 'a creativity or novelty in the world', almost a lawlessness *out of which laws can arrive.* I will return to this tantalizing view at the end of the book.

•

One little-noted but actually rather influential 'interpretation' of quantum mechanics was summarized by Asher Peres and Chris Fuchs in the title of an article they published in 2000: 'Quantum mechanics needs no interpretation'. Some researchers insist that quantum mechanics is already fully solved: that there are no interpretational difficulties remaining, no ambiguities or foundational matters waiting to be settled. This position demands that we simply accept certain givens – not because these are essential inputs into the theory, but because they lie outside of what quantum theory can be expected to explain.

Specifically, we must accept that *events exist* and that they happen *with a particular probability*: as Einstein might have put it, that God plays dice but that the dice eventually come to rest with one face uppermost. Then, quantum-mechanical predictions of probabilities are as good as it gets. If quantum mechanics says that an event will happen with a particular probability, there is nothing we can add to the theory to increase our certainty of its occurrence – nothing, that is, that will not get us into trouble elsewhere. Sure, you can ask questions about what is 'really going on', or about the mind–body problem or free will – but these are issues for philosophy, not physics. 'We could leave it at that', says the physicist Berthold-Georg Englert, an advocate of this 'completeness' viewpoint,

> were there not the widespread habit of the debaters to endow the mathematical symbols of the formalism with more meaning than they have. In particular, there is a shared desire to regard the Schrödinger wave function as a physical object itself after forgetting, or refusing to accept, that it is merely a mathematical tool that we use for a description of the physical object.

But not many share Englert's confidence that debates and arguments over quantum theory are 'diligent effort wasted on studying pseudo-problems'. And much of the reason lies in Englert's own remark. Yes, the Schrödinger equation is a tool for describing the physical object. But our very need to speak of a 'physical object' in the first place highlights the problem – for we can't then help but wonder about the nature of that object. One answer is to refer the question back to the Schrödinger equation itself: *this* is all that can be said about the object. Yes, but it is just a tool for describing it. Surely we are permitted to look beyond the mathematical tool to the object itself? Ah, but where does that get us? And so on.

As Fuchs and Peres put it, all we can ask of science is that it furnish us with a theory that can make predictions to test against experiment. If it can also supply a model of some free-standing 'reality', then so much the better. But, they say,

> There is no logical necessity for such a realistic world view to always be obtainable. If the world is such that we can never identify a reality independent of our experimental activity, then we must be prepared for that, too.

Yet even if we accept this limitation, it is by no means clear that we have already wrung from quantum mechanics all the insight we can get. It seems unlikely that this ragbag of lucky guesses and clever tricks, this awkward theory of astonishing accuracy conjured from an ontologically enigmatic formalism, is the last word on the matter.

The experimental and theoretical investigations of quantum mechanics over the past several decades have not yet helped to winnow the list of interpretations. They might have even encouraged further proliferation; as David Mermin has wryly noted, new interpretations appear regularly but none ever disappears.

But these recent studies have sharpened the questions, and have focused attention in places where the likes of Bohr and Einstein did not look. What is impressive is that what Bohr and Einstein (more than any of their contemporaries) had to say remains relevant. For we can now see, more clearly than they could have been expected to, what it is that they were really arguing about.

Whatever the question,

the answer is 'Yes' (unless it's 'No')

Quantum mechanics might seem 'weird', but it is not illogical. It's just that it employs a new and unfamiliar logic. If you can grasp it – if you can accept that this is just how quantum mechanics works – then the quantum world may stop seeming weird and become just another place, with different customs and traditions and with its own beautiful internal consistency.

Quantum logic describes how we get from abstract mathematical states to observable, tangible measurements. Why should we even need rules for that, though? In the everyday world we just go directly from one to the other without guidelines to get us there. One of the properties describing the state of my cup is that it is green. When I make a 'measurement' of its colour – by which in this case I mean simply, when I look at it – this greenness is registered by me as 'being green'. It sounds ridiculous even to put this into words. All I'm saying is that because my cup is green, it looks green to me.

The trivial prescription for going from classical states to measurements and observations ('It's in a state with property X, so I measure X') is replaced by a decidedly non-trivial one in quantum mechanics. Remember that the state, as described by the wavefunction, contains all that can be known about the quantum system. So anything that can be measured is in there somewhere. If it is not in there then it can't be measured; it relinquishes any claim to an 'is'.

But how does 'can be' become 'is'?

•

Let's start by asking what the properties *are* that make up a quantum state. We (unless we are physicists, perhaps) usually regard the world as a collection of *things*: trees, people, air, stars and planets. These have particular properties: colour, weight, smell and so forth. Some of these properties might be a little vague (texture, say), and some might be rather complicated outcomes of the way an object interacts with its environment (like shininess).* But we can usually break them down by reducing the 'things' themselves (if not the Thing In Itself) to more fundamental things: identifying their material constitution as arrangements of atoms with ninety or so varieties, for example.

It's not obvious why any of the properties that things have at the everyday scale should remain meaningful properties at the microscopic scale. Some don't. Electrons don't have a colour, nor do they really have a definable size. But they *do* have other familiar properties: mass, velocity, energy and electric charge. By the same token, some properties appear at the microscale that don't have any significance (at least in our daily experience) in the macroworld. Quarks, the fundamental components of protons and neutrons, have a property *called* colour, but it doesn't have anything to do with 'colour' in the sense of red apples and green leaves. It's just a label to distinguish

* I'm being intentionally materialistic, and ignoring those qualities and abstractions that give the world human value. Quantum mechanics has nothing to say about them, and there would be no need to apologize for that if some scientists didn't keep talking grandiosely about Theories of Everything that are silent about the things which matter most to most of us.

different types of quark and the way they interact. Physicists could instead have called it 'flotch'; at any rate, there was no ready-made word for such a concept, so they borrowed one (for better or worse).

There's another quantum-scale property like this that lacks an everyday analogue. It is called *spin*. Like the 'colour' of quarks, this is a familiar word repurposed to unfamiliar ends. But in this case there's a good reason for the choice. It's worth looking into that reason, not only to justify what otherwise seems like unnecessarily confusing terminology but also to see why yet again the history of quantum theory is a story of a struggle against the temptation to reach for classical stories to tell about it.

Spin was first introduced simply as a way of 'labelling' electrons, without any notion of what characteristic the label referred to. Niels Bohr's proposal in 1913 that electrons in atoms have quantized energies included the stipulation that these energy states are arranged into 'shells': loosely speaking, electrons in successive shells have increasingly greater energy. The shells, it turns out, have a substructure: each contains different kinds of electron orbits – more strictly *orbitals*, since, as we saw, the electrons don't orbit the nucleus in the conventional sense that applies to moons and planets. So an electron can be labelled according to its shell and orbital type, and also according to which specific orbital it occupies within a group of equivalent orbitals. These three labels are called quantum numbers, and they are literally just numbers: for example, the first shell of electrons has a quantum 'shell' number (denoted n) of 1.

In 1924 Wolfgang Pauli argued that electrons in atoms must be assigned a fourth quantum number too. He said that the characteristic features of atomic spectra – the way atoms absorb or radiate light due to electrons jumping between different energy states – can be explained if we assume that each electron orbital can accommodate precisely two electrons and no more. With just three quantum numbers, specifying an electron's shell, orbital type, and specific orbital within that family, the pairs of electrons within each orbital would be identically labelled: there's nothing else to tell them apart. But with a fourth quantum number that labels the two members of the pair in different ways, each electron in an atom then has a unique 'barcode' of quantum numbers, which together specify its quantum state. Pauli proposed that no two electrons can occupy the *same* quantum state: each *has* to have a unique set of quantum numbers in any given atom.

Pauli recognized that this rule could account for the structure of the periodic table of the chemical elements. The elements are arranged into groups according to the gradual filling up of their electron shells and orbitals, two electrons to each orbital. As we track across the elements from left to right in the table, each variety of atom possesses one more electron than the last, and this electron occupies the next 'slot' of lowest energy. Each time a new shell is filled, the next element begins a new row of the periodic table. And the electron-holding capacity of these shells and subshells determines the organized structure – before Pauli's principle, just a mysterious empirical fact – of the periodic table.

This is how things work in physics: you really can just make up a property in order to fit with what you observe, and only worry later about what that property corresponds to. So what property of an electron *was* Pauli's new

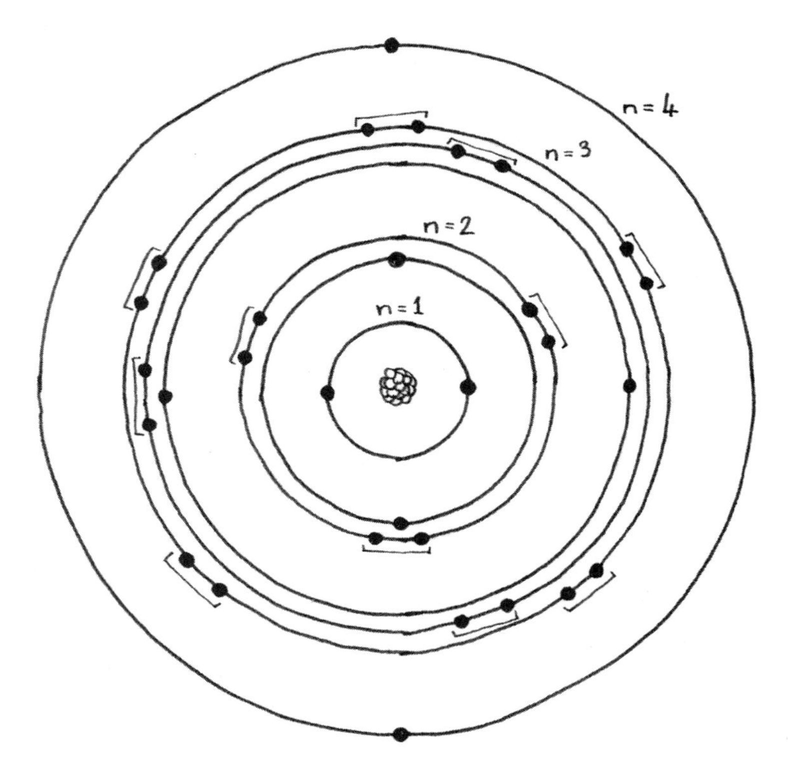

A crude representation of the arrangements of electrons in atoms. They are organized into shells around the central nucleus, and each shell is labelled with a quantum number (*n*). Each shell has a substructure of distinct orbitals (labelled with two other quantum numbers), and each orbital can accommodate two electrons, here shown as bracketed pairs. The fourth quantum number (spin) distinguishes the two electrons paired in each orbital. This image is highly schematic – the electrons do not really have circular orbits, but are distributed in space in more complex shapes. This image shows the electron configuration in the element zinc.

quantum number referring to? An answer was offered the following year by the Dutch physicists George Uhlenbeck and Samuel Goudsmit. Uhlenbeck proposed that electrons might be rotating around an axis like spinning tops, and that Pauli's fourth quantum number denoted whether the electron was spinning clockwise or anticlockwise. He and Goudsmit sketched out their idea in a short paper in 1925.

The same idea had occurred earlier that year to a young German–American scientist working in Germany, named Ralph Kronig. But Pauli had been so dismissive of Kronig's concept of a spinning electron that Kronig never tried to publish it. Only after the Uhlenbeck–Goudsmit paper started to excite discussion did Kronig appreciate that he should have had the courage of his convictions. (It took a great deal of courage to contradict Pauli, who, while still a young man himself, had a fearsome reputation both for his intellect and his scathing tongue.)

If the electron were a spinning ball of charge, then it would be magnetic. This follows from Michael Faraday's discovery in the early nineteenth century that electricity and magnetism are entwined. It's thanks to this relationship that an alternating electric current induces a magnetic force that spins the electromagnetic motor, and conversely that a magnet spun by some force, such as is exerted by flowing water or air, can generate an electrical current in a turbine.

Well, some entire atoms *are* magnetic. Two German physicists, Otto Stern and Walter Gerlach, found in 1922 that atoms may possess magnetic poles – the technical term is a magnetic moment, where here 'moment' is used with the meaning it has in the theory of mechanics, being a force that induces rotation. If such atoms are propelled between two magnets, so that they pass through a north–south magnetic field, they experience a magnetic force depending on how the atoms' magnetic poles are oriented relative to the applied magnetic field. This force may deflect them from their initial paths.

After the notion of electron spin became established, it was quickly understood that the Stern–Gerlach experiment can be interpreted in terms of the magnetism of the atoms' electrons: an atom's total magnetism is the sum of the electrons' magnetic moments. Each orbital may contain two electrons, and if it does then their spins are opposite. Then their magnetic moments point in opposite directions and cancel out. But if the atoms contain orbitals with unpaired electrons, they can contribute to the net magnetic moment of the atom. It later became clear that atomic nuclei too can have magnetic moments, depending on their particular combination of protons and neutrons (which also have spin). An atom's total magnetic moment is a combination of those of the electrons and of the nucleus.

It made sense that electrons should have a magnetic moment, if indeed they are charged particles that *spin*. However, Stern and Gerlach discovered that the magnetic moment of the electron is *quantized* – it can only take particular values. Specifically, it can only have a single magnitude, never bigger or smaller than that. This was no great surprise, for by then quantization was recognized as a fundamental property of such particles: that's what quantum theory seemed initially to be all about. The natural way to understand the quantization of the electron's magnetic moment was that its *spin* can only have two values: the particle can spin at a specific rate, and no other, either in one direction or the opposite.

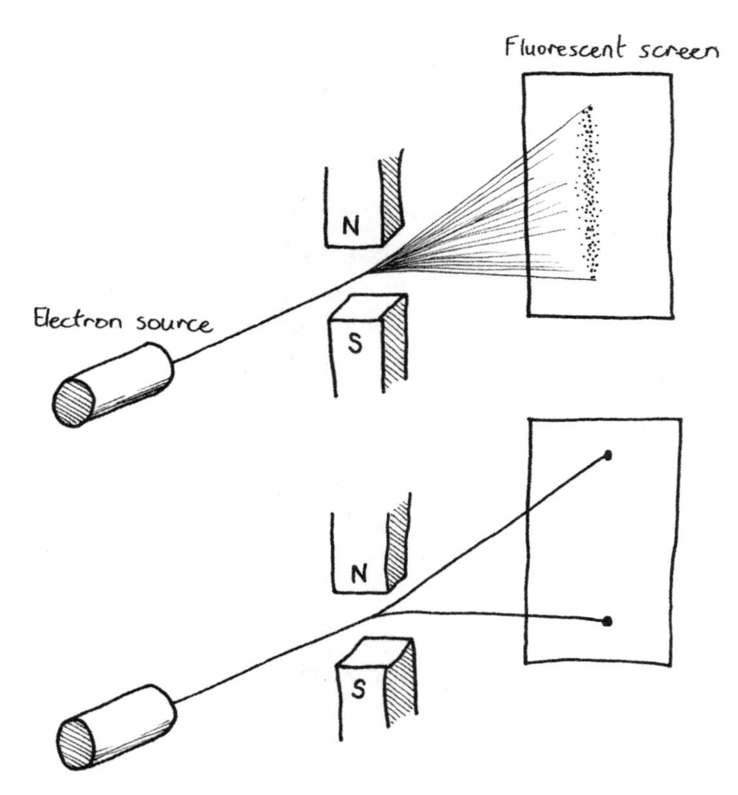

The Stern–Gerlach experiment for electrons. If their spin, and therefore their magnetic moment, can take any value, one would expect a range of deflections due to the magnetic field (top). But in practice the electron beam is only deflected by a fixed amount in one direction or the other, because the spins are quantized and may take only one of two possible values, with opposite orientation (bottom).

This all adds up to a clear intuitive picture: the electron spins with quantized rotational motion, and by virtue of its electrical charge this makes it magnetic.

There's just one problem.

The rate of spinning is related to a quantity called the angular momentum, which is simply the momentum of a rotating object. An object travelling in a straight line has linear momentum, a quantity equal to the object's mass multiplied by its velocity. Similarly, angular momentum is related to the mass of the rotating object times its rotational speed. But when we consider how this relationship for the spinning electron connects to its magnetic moment, we find something odd. If we analyse the relationship using classical mechanics and electromagnetic theory, we seem forced to conclude that the electron has to rotate *twice* to get back to where it started.

That's a meaningless statement really. The very definition of a complete rotation is that it restores the object to where it started from. We can't picture what it can mean to say that a complete rotation only gets you 'halfway' there.

What this *does* mean is that the electron is not spinning in the normal sense. It isn't clear *what* it is 'doing' to produce its magnetic moment. Classical physics gives us a sense of how magnetism can be generated by rotation of a charged object, but that can't be quite what's going on for the electron. There is no everyday picture we can use here. So if you read somewhere (as one sometimes does) that the quantum spin of an electron is weird because two complete revolutions are needed for a 'full turn', don't take too much notice. We just don't know how to picture quantum spin. It is some property that makes the particle respond, like a magnet, to an external magnetic field, and that's all. There is *no* classical analogue. Even though in some respects quantum spin does resemble the effects that classical spin induces, 'any attempt to visualize it classically will badly miss the point', according to Leonard Susskind.

Spin has a deep significance in quantum theory. It turns out that there are two distinct classes of fundamental particle: those with spin quantum numbers that are integers (0, 1, 2 ...) and those with half-integer spin (½).* The former are called bosons, and they constitute all the 'carrier' particles of fundamental forces: gluons (carrying the strong nuclear force that binds together the constituents of atomic nuclei), so-called W and Z mesons (carrying the weak nuclear force involved in radioactive beta decay) and photons (carrying the electromagnetic force). You will probably have heard also of the Higgs boson, which is involved in the way particles acquire some of their mass and is unique in being the only fundamental particle so far known that has zero spin. The particles with half-integer spin, meanwhile, are called fermions, and they are what make up everyday matter: electrons, protons, neutrons (those latter two are composites of fermions called quarks) and others.

So spin divides fundamental particles into two groups. Why this should be so, no one knows. But many physicists hope that, as we probe beyond the Standard Model of particle physics using instruments such as the Large Hadron Collider at the CERN particle-physics centre in Geneva, we will discover an underlying relationship between bosons and fermions, perhaps via a new principle of physics called supersymmetry.

* Fundamental particles with spin 3/2, 5/2 and so on are possible in principle, but have never been observed. Some theories predict spin-3/2 particles as candidates for the mysterious 'dark matter' thought to make up about four-fifths of the mass of the universe, or as components of a (still elusive) quantum theory of gravity.

•

When I say spin is quantized, I mean that a measurement of the magnetic moment it induces can only ever deliver one value of its magnitude, in two possible orientations. That's to say, it can have only one 'size', but pointing in opposite directions. We can loosely call these 'spin up' and 'spin down'. This sounds straightforward, perhaps – after all, isn't this discrete quantization what quantum mechanics is all about? But quantization doesn't exactly mean, as is often implied, that quantum quantities are obliged to take certain values and no others. The quantization is not so much a property of the system we are studying; it is a property of measurements we make on it.

Say that we're going to make measurements of the spin of electrons. We know already – because I've told you this is so – that the *size* of the spin will be ½ (don't worry about the units). We just want to measure its direction.

Let's define our frame of reference: our grid of spatial directions. I'll call up–down the *z* direction, and the two horizontal directions (north–south and east–west, you could say) are *x* and *y*. In general the electron spin could point in any direction: *x*, *y*, *z* or anything in between. We can break down this spin conceptually into three components, one in each direction, which we'll label σ_x, σ_y and σ_z. If we imagine the spin as being like a flagpole pointing in some direction in space, these components are like the length of the shadow the pole makes when illuminated from each respective direction. If the spin points up in the *z* direction, say, then $\sigma_z = +½$, while σ_x and σ_y are zero.

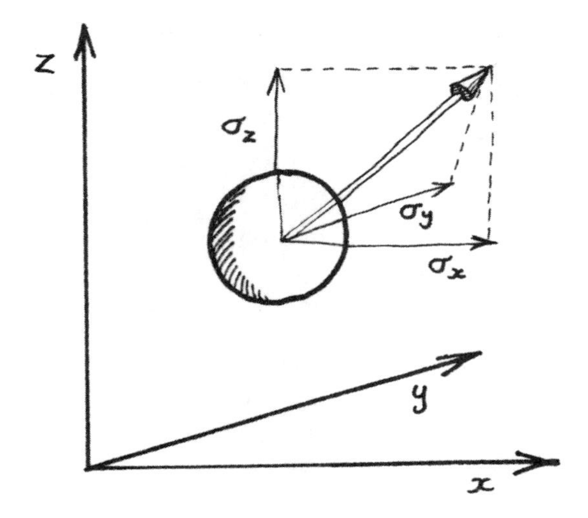

This is how we might think classically about the spin components of an electron: the spin is in some arbitrary orientation, and the three components σ_x, σ_y and σ_z are projections of that spin onto the three axes of space. But it is not clear that we can assign any physical reality to such a picture, since measurement of a spin component can only ever deliver a value equal to the quantum of spin: ±½, regardless of which orientation your measuring device is in.

We can 'prepare' the electrons in a specific spin state by orienting their spins in a magnetic field. It's rather like orienting the magnetic moments of iron atoms all to point the same way by stroking a needle with a magnet. Let's say we do indeed align them 'up' in the *z* direction. If we now measure σ_z for one of these oriented electrons, we expect to find that it has the value +½.

And we do! Yes, sometimes in quantum experiments you know *just* what you're going to get (because you've set things up that way).

A simple way to measure the *z* component of the electrons' spin is to perform a Stern–Gerlach experiment. We direct the electron beam into the gap between two magnets oriented so that the field is in the *z* direction. The spin-oriented electrons will be deflected by the field, all in the same direction: the beam will be bent. If, in contrast, we had prepared the electrons with their spins *randomly* oriented in the *z* direction, we'd see the beam split neatly in two: the 'up-spin' (+½) electrons would be deflected one way, the 'down-spin' (−½) electrons by the same amount in the other direction.

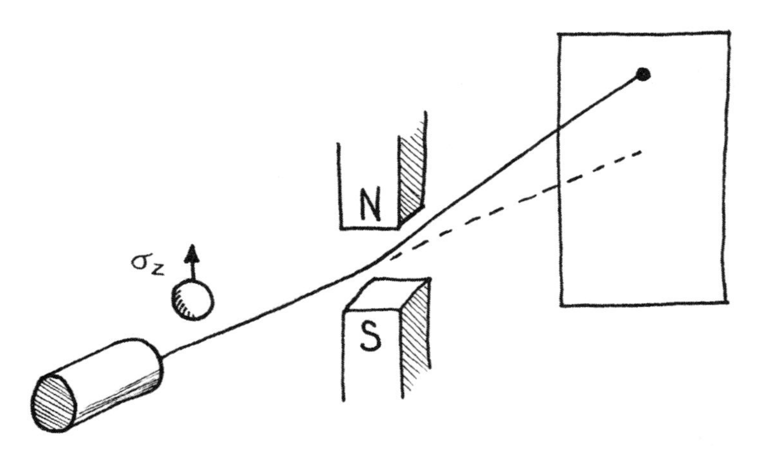

Measuring σ_z for spins that have been prepared with σ_z +½ yields the expected result: in this Stern–Gerlach experiment, a uniform upward displacement of the beam.

Now let's ask what is the value of the *x* component of the spin (σ_x). It should be zero, right? Because we prepared the electrons with their spins oriented in the *z* direction, the 'shadow' projected in the *x* direction has zero 'length'.

Let's try it. We take the deflected particle beam from our initial measurement of σ_z and pass it between two magnets oriented to produce a horizontal magnetic field along the *x* direction: a second Stern–Gerlach measurement. If σ_x = 0, there should be no deflection. But there is: the beam is split in two, indicating that in fact σ_x for each electron has the value +½ or –½, and that these proportions are equal on average – in other words, a random mixture of the two.

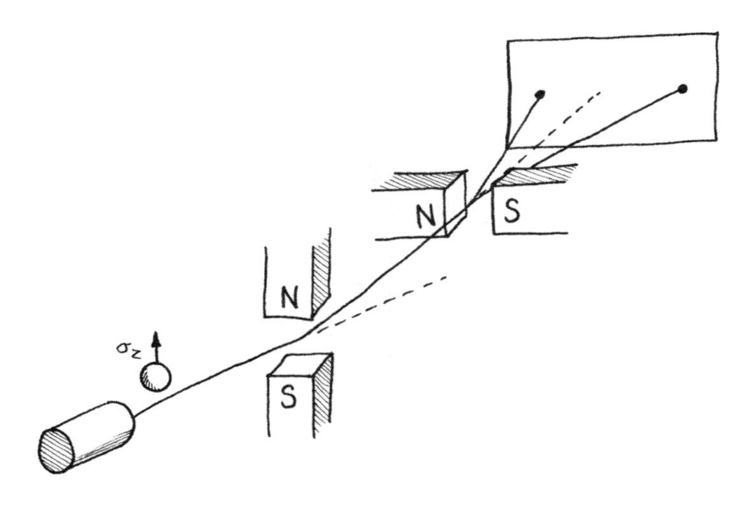

Measurements of spin can only take the value ±½. So even if spins have been prepared with components σ_z = +½ and σ_x = 0, a measurement of σ_x will elicit the two values σ_x = ±½ in equal measure - with the *average* value of zero.

What happened? This outcome is a simple consequence of quantization. If we measure *any* component of these particles' spins, it must take the value ±½, because those are the *only values permitted* – that's exactly what quantization means. All the same, the 'classical' expectation is satisfied *on average*. If we do the experiment for any individual electron, we measure first $\sigma_z = +\frac{1}{2}$ (as we expected) and then subsequently $\sigma_x = +\frac{1}{2}$ or $-\frac{1}{2}$ at random. So if we keep doing this experiment over and over again we will find that the average value of σ_x over many measurements is indeed zero.

What we expect from the classical analogy is therefore verified by repeated measurements, but contradicted by any individual measurement. It's as if the apparatus 'knows' that, because we've initially oriented the spins in the *z* direction, the *x* (and *y*) component 'ought' to be zero – but it can't deliver that value, because quantization implies that any measurement of spin can only see the values $\pm\frac{1}{2}$. So it does the next best thing, contriving to make the average value of these components zero instead.

If a measurement of σ_x gives values of $\pm\frac{1}{2}$ at random, what happens if we now measure σ_z a second time, in a third Stern–Gerlach experiment with appropriately oriented magnets? We find that now it too has a value of $\pm\frac{1}{2}$ *at random*. Measuring σ_x scrambles σ_z. That still holds if we skip the first measurement of σ_z and just measure σ_x followed by σ_z. In other words, having first prepared the particle in the $\sigma_z = +\frac{1}{2}$ state gives a 100% chance of finding $\sigma_z = +\frac{1}{2}$ if we measure it before we measure anything else. But a measurement of first σ_x and then σ_z gives only a 50% chance of that outcome, despite the fact that the initial state is the same. The order of measurement matters.

Well, maybe *any* measurement of spin scrambles it, so that all subsequent measurements will give random results? Let's check by measuring σ_z twice in succession. No, that's fine: the electron spins that were oriented in the first measurement ($+\frac{1}{2}$) stay oriented in the second (still $+\frac{1}{2}$).

It looks, then, as though measuring σ_x specifically (and we'd find that the same applies to measuring σ_y) for a particle prepared with $\sigma_z = +\frac{1}{2}$ will scramble the spin's orientation. But that doesn't make much sense. You're doing the same experiment in every case: a Stern–Gerlach measurement of spin. Why should one such measurement (in the *x* direction) scramble the spin orientation while the other – identical except for an arbitrary change of orientation to the *z* direction – does not? It's as if the apparatus 'knows' which answer you *should* get (on average) and therefore 'knows' whether or not to scramble the spin orientation during the measurement.

Compared with the classical world, there seems to be a different logic at work here. One of the strangest implications of this quantum logic is that, as we just saw, it can matter in what order measurements are performed. Measuring σ_z and then σ_x gives different outcomes from measuring σ_x and then σ_z. The consequences of this non-equivalence of the ordering are profound.

Not everything is

knowable at once

If there's one thing most people know about quantum physics, it's that it is uncertain. There is (we're told) a fuzziness to the quantum world that prevents us from knowing everything about it in absolute detail. Ninety years ago, Werner Heisenberg articulated this with his famous Uncertainty Principle.

But Heisenberg's discovery is often misunderstood. It might be taken to imply either that nothing in the quantum world can be measured exactly (perhaps because we can't help disturbing what it is we want to measure?) or – a more sophisticated misconception – that if we want to measure one thing very accurately then we have to accept a commensurate blurring in the values of everything else. Neither idea is correct.

We can't blame these misunderstandings on a woeful lack of public science literacy. The Uncertainty Principle actually is rather technical, and it's not surprising that non-specialists may miss its message. The problem is compounded by the catchy name. It's a name that chimed with the insecure times in which Heisenberg deduced his result in 1927: between the wars, with Germany reeling from hyperinflation and political crises, and with Nazism on the rise. To make things worse, even Heisenberg did not fully understand the implications of what he had stumbled across. He couched the Uncertainty Principle in terms both hazy and apt to mislead, and which have left physicists arguing about it even today. He made trouble for himself.

Heisenberg's Uncertainty Principle is not exactly a constraint on how precisely we can make a measurement of some quantum property. Rather, it constrains how precisely the property we want to know about exists at all. It might have been better christened the Unknowability Principle – better still, the Un*be*ability Principle – although doubtless that would have spawned a mysticism of its own.*

The point is this. Quantum objects may in principle have a number of observable properties, but we can't gather them all (Copenhagenists might in fact say 'elicit them') in a single go, *because they can't all exist at once*. And by gathering some we may scramble the values of others. We've just seen how that applies to the spatial components of the quantum property called spin. I didn't say so at the time, but the scrambling of spin components, and the significance of the order in which we choose to measure them, arise precisely because the mutual relationships between these components are governed by the Uncertainty Principle.

Now let's see what 'uncertainty' really means.

•

It wasn't any failed attempt to make accurate measurements that motivated Heisenberg's paper on the Uncertainty Principle. He was, after all, a theorist (what is more, one with a somewhat shaky grasp of experimental

* Heisenberg of course used German words: in his original 1927 paper he refers to both *Ungenauigkeit* (inexactness) and *Unbestimmtheit* (undeterminedness or vagueness). 'Undeterminedness' is closer to the mark, but translation is itself necessarily inexact. 'Uncertainty' is perhaps closest to Niels Bohr's preferred term *Unsicherheit* (doubtfulness or unsureness). Here we could fairly accuse Bohr of forgetting his usually scrupulous way with words.

methods). Like most of his contemporaries, Heisenberg was trying to understand the quantum world by developing a mathematical formalism that could capture what little was then known about experimental quantum physics (such as the way atoms absorb and emit light), and then to see where it led. Often it led to 'thought experiments' that no one knew how to perform. It was a hugely abstract and intellectual exercise, relying on informed guesswork to an extent that we might consider both impressive and alarming.

So the Uncertainty Principle was purely a mathematical deduction. Heisenberg was saying that, if we have deduced the correct logic of quantum mechanics, then there is a strange corollary. Suppose we want to know what values two properties of a quantum system, p and q, have. We devise some experiment for measuring them both. Now, there is always some error and uncertainty in making a measurement, because of the limitations of your apparatus – that's true in classical physics too. But as techniques improve, so does accuracy. Yet the Uncertainty Principle says there's a limit to this improvement, in the following sense. As we get better at measuring p, we find that there's a limit to how precisely we can *at the same time* pin down q. There is an unavoidable trade-off between how precisely we can measure p (let's denote the imprecision Δp) and how precisely we can measure q (Δq). Specifically, the product Δp times Δq can never be less than the value $h/2\pi$. Here π has its usual geometric meaning of the ratio of the circumference to the diameter of a circle, and h is the fundamental constant called Planck's constant, which sets the scale of the 'granularity' of the quantum world – the size of the 'chunks' into which energy is divided (page

28). This h is extremely small, so the Uncertainty Principle only matters once we have got very accurate indeed about measuring p and q. But we can't ever know the two of them as precisely as we like at the same time.

The Uncertainty Principle applies, for example, if p is the momentum of an object (its mass times its velocity) and q is its position. This limitation is conveyed in one of the many half-baked jokes about the Uncertainty Principle:

> Heisenberg is pulled over for speeding. The police officer asks him, 'Do you know how fast you were going?'
>
> 'No,' Heisenberg replies, 'but we know exactly where we are!'
>
> The officer gives him a confused look and says, 'You were going 108 miles per hour!'
>
> Heisenberg throws his arms up and cries, 'Oh great! Now I'm lost!'

As with many science jokes, any residuum of amusement must be expunged if we want to be accurate. It's not that *any* measurement of speed (more precisely, momentum) renders position completely uncertain; rather, the more accurately the speed is known, the more uncertain the position.

But more importantly, the joke illustrates exactly what it is that is generally misunderstood about the Uncertainty Principle. Heisenberg is obviously *somewhere*; he just doesn't know where (which is surely the definition of 'lost'). A more rigorous statement of the situation, however, is that if Heisenberg's speed is known to within a certain degree of accuracy, his position is *undefined* with a scope defined by the uncertainty relationship above. He can only be *said* to have a position at all to within these bounds. Which is even less funny.

What's more, this restriction on precise knowledge does *not* apply to all pairs of quantum properties. It applies only to some, which are said to be 'conjugate variables'. Position and momentum are conjugate variables, and so are energy and time (although the uncertainty relationship between them is subtly different from that between position and momentum). It doesn't apply to, say, the mass and the charge of a particle: we can measure these simultaneously as accurately as we like. I have never found an intuitive explanation of what makes two variables conjugate (although one can certainly express that in formal mathematics). What we can say, however, is that this 'uncertainty' has a far more exact meaning than that the quantum world is all a bit fuzzy.

•

How did Heisenberg figure out that some variables are related this way, if not by actually observing it? His Uncertainty Principle falls out of the math. There are various ways of deriving it, but perhaps the most instructive makes reference to Heisenberg's own mathematical formulation of quantum mechanics, called matrix mechanics. This was a rival to Schrödinger's 'wave mechanics', although the two descriptions are in fact equivalent; for many purposes, Schrödinger's is simply easier to use. Thinking about the Uncertainty Principle using matrix mechanics is instructive because it shows that this is not some weird behaviour conjured mysteriously by the 'otherness' of the quantum world, but is an implication of its mathematical logic, which can be understood in terms of school-level math.

Heisenberg's matrices were tabulations of the quantum properties of objects – basically a way to write down quantum states, along with the operators that describe transformations of these states and from which we can predict measurable quantities. There's a well-established form of arithmetic for working with matrices, and all you need to know about it is that it's not like the ordinary arithmetic of pure numbers. If we multiply together two numbers, it doesn't matter in which order we do it: 3×2 is the same as 2×3. For matrices this is no longer true. If we have two matrices **M** and **N**, then $\mathbf{M} \times \mathbf{N}$ is not necessarily the same as $\mathbf{N} \times \mathbf{M}$. In other words, the difference $\mathbf{M} \times \mathbf{N}$ minus $\mathbf{N} \times \mathbf{M}$ is not necessarily zero. The order in which we do the math matters.

This property is known as *non-commutation*. What Heisenberg realised is that, in quantum matrix mechanics, the operators that reveal certain properties of a quantum state don't commute. This is the feature that makes those properties conjugate variables. Heisenberg showed that the difference between performing the successive operations in one order and in the other order is equal to the accuracy margin of the Uncertainty Principle: $h/2\pi$.

That we should get different results from conducting two operations – which may correspond in the real world to making two measurements – in a different order could seem strange. It is just what we found earlier for the measurements of spin components: if we measure one first, we might scramble the other (which is to say, we render it 'uncertain'). But it's not such an unfamiliar notion really.

Cookery supplies the popular metaphor: if we add the baking powder *after* we have mixed the other ingredients and baked the cake, we get a different result from adding it *before* baking (I have tried this experiment, so I can verify the statement). I like to think that a better, if peculiarly British, analogy is supplied by making tea. Adding the milk to the cup *before* pouring the tea gives a different quality of brew to adding the milk *to* the tea in the cup. It might seem improbable, but connoisseurs swear it is true (I believe them), and there may be good scientific reasons for it. (They are not, I should add, quantum reasons.)

•

Yet surely it is no explanation of the Uncertainty Principle to say that it comes from non-commutation in the math! Granted, it is what follows if we accept quantum math – but the Uncertainty Principle is also something we can observe in real experiments. Heisenberg and his contemporaries couldn't do that, but instrumental techniques are now good enough to show us this effect in action: a blurring of resolution in one conjugate variable if we get more accurate about observing the other. And when we seek explanations for things we can measure, it doesn't seem very satisfying to point to equations. We want some physical picture of what is going on. What is it that causes the blurring?

Surprisingly, given his general disdain for attempts to visualize the formal principles of quantum mechanics, Heisenberg seems to have felt obliged to answer that question. The paper in which he presented the Uncertainty Principle even advertised this in the title: 'On the visualizable content of quantum theoretical kinematics and mechanics'. Heisenberg might, however, have been better advised to heed his normal aversion, because the physical picture that he offered was misleading and continues to muddy the waters today.

The imprecision in our ability to measure two properties at once, Heisenberg suggested, comes from the smallness and delicacy of a quantum particle. It is virtually impossible to make a measurement on such an object without disturbing and altering what we're trying to measure. If we 'look' at an electron by bouncing a photon of light off it in a microscope, that collision will change the path of the electron.* The more we try to reduce the intrinsic inaccuracy or 'error' of the measurement, say by using a brighter beam of photons, the greater the disturbance. According to Heisenberg, error (Δe) and disturbance (Δd) are also related by an uncertainty principle, according to which the product $\Delta e \times \Delta d$ again can't be smaller than $h/2\pi$.

The American physicist Earle Hesse Kennard showed soon after Heisenberg's paper was published that this gamma-ray-microscope thought experiment is superfluous to the issue of uncertainty in quantum theory. The restriction on precise knowledge of both speed (more properly, momentum) and position is an intrinsic

* Heisenberg rightly understood that to 'see' such small objects we would need photons of extremely short wavelength, such as gamma rays. However, his ignorance about the basic physics of microscopes had almost led him to fail his doctoral examination in 1923, and it hadn't improved much four years later. When he presented his 'gamma-ray microscope' picture to Bohr, his Danish mentor was obliged to correct some of the misconceptions in Heisenberg's argument.

property of quantum particles, not a consequence of the limitations of experiments.

There *is* a more 'physical' way of understanding the Uncertainty Principle, rather than seeking justification in the obscure non-commutation of matrices. It comes from the idea that quantum particles can show both a wave-like nature, spread out in space, and a localized particle nature. To get a good probability of finding the particle in a small region of space from a wave-like probability distribution, we can combine waves of different wavelengths such that they interfere constructively (page 66) in just that region but destructively everywhere else. This localized wave is called a wave packet. To increase the localization and get a more tightly defined position for the particle, we must add more waves. But the wavelength determines the particle's momentum. So the more waves there are, the more possibilities there are for a measurement of momentum.

Heisenberg's thought experiment shows that he had still not fully grasped what Bohr was saying about quantum mechanics. The 'disturbance' view implies that the particle being observed really does have some precisely defined position and momentum, but that we simply can't measure these things accurately without changing them. That difficulty would then presumably apply to any quantum property, not just to conjugate variables. But for Bohr, all one can meaningfully say about a quantum system is contained in the Schrödinger equation. So if the math says that we can't measure some observable quantity with more than a certain degree of precision, that quantity simply does not exist with greater precision. That is the difference between uncertainty ('I'm not sure what it is') and unknowability ('It *is* only to *this* degree').

•

Heisenberg's 'experimental' version of the Uncertainty Principle – his relationship between error and disturbance – has nevertheless continued to interest physicists. It seemed as though one really could use the Uncertainty Principle to derive some general relationship between these things: a relationship constrained by the condition that $\Delta e \times \Delta d$ can be no smaller than $h/2\pi$.

Recently that notion has become the subject of intense debate. In 2003 the Japanese physicist Masanao Ozawa argued that it should be possible to defeat Heisenberg's apparent limit on error and disturbance. He proposed a new relationship between these two quantities in which two other terms were added to the equation. In other words, $\Delta e \times \Delta d + A + B$ (never mind exactly what A and B are) could be no smaller than $h/2\pi$, so that $\Delta e \times \Delta d$ itself *could* be smaller than that. Ozawa's new relationship has now been tested in two separate experiments involving beams of photons and neutrons. Both experiments showed that the precision of the measurements could indeed violate Heisenberg's limit on $\Delta e \times \Delta d$, but not Ozawa's.

Some other researchers have taken issue with these claims, but it all seems to depend on exactly how you pose the question. Heisenberg's limit on how small the combined uncertainty can be for error and disturbance holds true if you think about averages of many measurements, but Ozawa's smaller limit applies if you think about individual measurements on particular quantum states. In the first case you're effectively measuring something like the 'disturbing power' of a specific instrument; in the second case you're quantifying how much we can know about an individual state. So whether Heisenberg was right or not depends on what you think he meant (and perhaps on whether you think he even recognized the difference).

The debate highlights how quantum theory does not ascribe some generalized fuzziness in the microscopic world. Rather, what the theory can tell you depends on what exactly you want to know and how you intend to find out about it. It suggests that 'quantum uncertainty' isn't a sort of resolution limit, like the point at which objects in a microscope look blurry, but is to some degree chosen by the experimenter.

This fits well with the emerging view of quantum theory as, at root, a theory about *information* and how to access it. Recent theoretical work by Ozawa and his collaborators suggests that the error-disturbance relationship is a consequence of how gaining information about one property of a quantum system degrades the information we can possess about other properties of that system. It's a little like saying that you begin with a box that you know is red and think weighs one kilogram – but if you want to check that weight exactly, you weaken the link to redness, so that you can't any longer say for sure that the box you're weighing is a red one. The weight and the colour become interdependent pieces of information about the box.

This is hard to intuit, I know. But it's a reflection of how interpretations of quantum theory are starting to shift towards the view that what we can know about the world depends not on some fundamental uncertainty or constraint, but on how we ask.

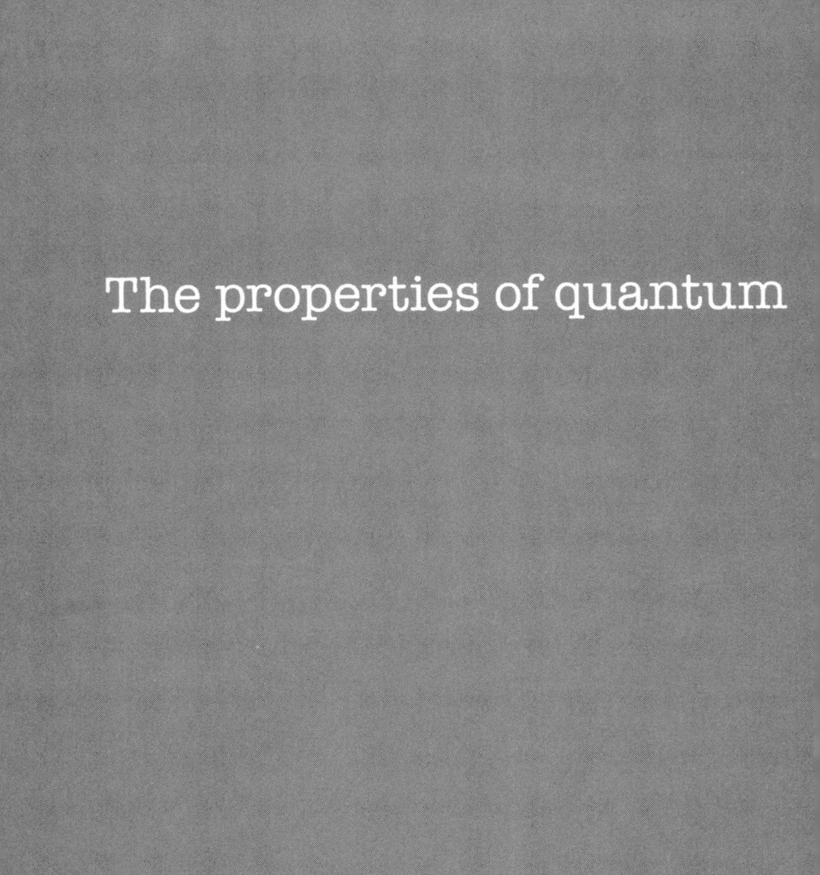
The properties of quantum

objects don't have to be
contained within the objects

Albert Einstein seemed resigned with good grace to a widespread determination to prove him wrong. An endless stream of cranks has attempted to 'disprove' Einstein's theories of relativity ever since they were first published, and Einstein responded patiently to some of the untutored correspondence he received that claimed to find errors in his work. Obviously, if you could show that Einstein had erred then you would be revealed as a genius of the highest degree, and there was (and still is) no shortage of applications for that position.

It is a sign of supreme intellectual renown when even your 'errors' and 'blunders' are celebrated, and when announcements that you have been proved 'wrong' make newspaper headlines. But actually Einstein was 'wrong' about many things. He made a few trivial lapses in his calculations. He famously fudged his theory of general relativity to avoid its prediction of an expanding universe, just a few years before astronomers found that to be precisely the state of the cosmos. Even his many proofs of the celebrated $E = mc^2$ contained little gaffes. Heck, there's an entire book enumerating Einstein's mistakes.*

None of this has the slightest bearing on Einstein's status as the greatest scientist of the twentieth century. To imagine that genius implies freedom from error is to misunderstand the nature of creativity and insight.

* It's *Einstein's Mistakes: The Human Failings of Genius*, by Hans C. Ohanian. *Einstein's Greatest Mistake* by David Bodanis is a more measured biographical account of his later life.

Arguably, geniuses (whatever that means) must incur an above-average chance of being wrong.

A favourite 'error' with which we delight in saddling Einstein is his failure to embrace the implications of quantum mechanics. No doubt this is partly because he expressed his scepticism in such a memorable sound bite: 'God does not play dice with the universe.' Perhaps too it might seem reassuring to imagine that even Einstein was unable to take an imaginative leap beyond the preconceptions of his early training. The notion that randomness, indeed absence of causality, lies at the heart of things is unsettling in the extreme, and there's some comfort in seeing that Einstein shared our instinctive reluctance to countenance it.

Yet it is a tired and muddled cliché that makes Einstein the stick-in-the-mud who could not accept the quantum theory he did so much to initiate. His intellectual sparring partner Niels Bohr was often frustrated and perplexed by Einstein's resistance to the new ideas, but Bohr would never have branded Einstein a stubborn conservative. His challenges undoubtedly helped Bohr to formulate and refine the framework on which he hung his own interpretation of quantum mechanics, and Einstein's opposition to the Copenhagen Interpretation stemmed not from pig-headed denial but from an unusually clear-sighted appreciation of what Bohr was saying. If he had not understood quantum mechanics so well, Einstein would have been less troubled by it.

In fact, Einstein deserves much of the credit for discovering what is arguably the central feature of modern quantum theory: as Erwin Schrödinger saw it, '*the* characteristic trait of quantum mechanics, the one that enforces its entire departure from classical lines of thought'. Einstein described this trait in 1935, and in the same year Schrödinger gave it the name by which it is now known: *entanglement.*

If Einstein isn't always afforded due praise, however, that at least is understandable. For he 'discovered' entanglement via a thought experiment that, because it posed an apparent paradox, demonstrated in his view that such behaviour couldn't possibly be real. Einstein wanted to bury entanglement even as he unveiled it.

The emerging appreciation of entanglement's role in quantum mechanics over the past several decades has shifted the emphasis of the whole field. Entanglement is indeed a real attribute of quantum objects, as numerous careful experiments since the 1970s have demonstrated. These studies have been advertised time and again as evidence that 'Einstein was wrong' about quantum theory. Yet much of that discussion is, unfortunately, wrong about why Einstein was 'wrong'. Perhaps the headline writers should have heeded Bohr, who said that the opposite of a profound truth is also a profound truth.

•

The main thing you need to know about entanglement is this: it tells us that a quantum object may have properties that are not entirely located on that object.

At least, that's one way to say it. There's no unique way to express what entanglement is. Yet again we lack words to convey such concepts with precision and clarity, and so we need several different ways of looking at them before we can begin to grasp what they are about.

What can it mean to say that an object's properties are not located solely on that object? My pen is black – the blackness doesn't have any existence beyond the pen. But what if I were to say that the blackness of my pen is also partly associated with my pencil? I don't mean that the pencil is black too. I mean that the blackness of the pen is partly in the pencil.

It doesn't seem to mean a great deal, does it? Here is another way of looking at it. If my pen and pencil were entangled quantum objects, it could be that I might examine my pen and find out all that there is to know about it and still not know for sure what colour it is – because the colour is not entirely *there in my pen.*

Or I could investigate the pen and pencil together, measuring all that there is to know about the two of them as a pair of entangled objects. I can measure, let's say, what 'colours' they share between them. And yet if they are entangled then I might end up with complete knowledge – that is, everything knowable – about the pair while being able to say little – or possibly *nothing at all* – about what each individual element is like, such as what colour they are. This is not because I haven't looked closely enough. It is because the entangled pen and pencil may not have local properties. They can't be ascribed individual colours.

That is, roughly speaking, what entanglement is like. It is, you might say, a quantum phenomenon through which single objects may be deprived of a definable character. Let's see how it came into the picture.

•

What most disturbed Einstein about quantum mechanics – what he was driving at with those divine dice – was the replacement of causation with randomness. We saw that if we prepare a particle in a state with its vertical spin component oriented *up*, and then measure a horizontal component, we find that it will be *up* or *down* with exactly 50% probability of each. A sequence of measurements for particles prepared in identical ways might yield *up, up, down, up, down, down* and so on. Why, in any particular experiment, do we measure one value but not the other?

You might imagine that this sort of randomness is nothing special. When a pin is balanced vertically on its point and allowed to fall, the direction in which it topples is random. Or is it? Suppose we could measure its state before toppling with extreme accuracy. Then we might find that it wasn't perfectly upright to begin with, so we have introduced a bias that decides the direction. So we get a better method of alignment. But now we see that the pin itself isn't perfectly symmetrical: there's slightly more mass on one side, and so that is weighed down more. So we make the pin perfectly symmetrical. But now we can detect exactly how many air molecules strike it from different directions, and find that in each experiment there's a very tiny imbalance that pushes the pin one way or another. So we let it topple in a vacuum – and so on. The point is that what seemed random turns out in each case to have a definable cause that produced a bias. Apparent randomness was simply a consequence of our lack of knowledge about the system.

This kind of randomness is easy to accept, because we may rest assured that there's a causal logic to what we see even if we can't get at it: in short, things happen for a reason. But the probabilistic nature of the Schrödinger equation, which predicts only the likelihood of different experimental outcomes, leaves it offering no reason why one specific outcome is observed instead of another. In effect, it says that quantum events (the radioactive decay of an atom, say) happen for no reason. *They just happen.*

That sounds like a terribly unscientific thing to say, and seems to go against the grain of everything that scientists and natural philosophers have striven to achieve since well before the time of Isaac Newton: to explain the world. Quantum events don't appear to have an explanation as such – one in which definable causes lead to specific effects – but only a probability of occurrence. This is what Einstein found unreasonable. Who can pretend that it isn't?

He suspected that this apparent randomness is just like the randomness of a toppling pin: it really does have a specific, deterministic cause (*this* leads to *that*), but we can't see what that is. It looks as though the particle has just decided its spin orientation on a whim, in the instant, but actually that orientation was fixed all along – yet hidden from view. Or rather, some property of the particle predestines the outcome of the measurement. This obscured property that allegedly renders quantum mechanics deterministic became known as a *hidden variable*.

But if, by definition, hidden variables can't be seen, how can we know they exist? In 1935 Einstein, collaborating with two young theorists named Boris Podolsky and Nathan Rosen, described a thought experiment showing (they claimed) that without hidden variables – that is, if you accepted the Copenhagen Interpretation – you were faced with an impossibility: a paradox.

In the Einstein, Podolsky and Rosen (EPR) experiment, two particles are produced in a way that makes their quantum states interrelated – entangled, as we'd now say. Because of this correlation between their properties, a measurement made on one of them would provide instant information about the other one too. And that's the problem.

The original EPR experiment is a bit hard to visualize, or indeed even to understand. The paper in which it was described lacks Einstein's usual clarity, because it was written by the Russian Podolsky. Einstein attested that it didn't really reflect his own views on the matter. 'It has not come out as well as I really wanted', he wrote to Schrödinger.

But the EPR thought experiment was given a more transparent formulation in 1951 by David Bohm. Let's imagine, Bohm said, that we create correlations between properties of quantum particles that have a discrete set of possible measured values: spins being *up* or *down*, say, or photons of light polarized vertically and horizontally. We set up a correlation whereby, if one particle has one of the two permitted values (spin *up*, say), the other has the other value (spin *down*). Researchers already knew by that time how to create photons with correlated polarizations of this sort: one can stimulate atoms with energy in such a way that they will emit two photons at the same time, entangled in this fashion. So Bohm's version of the EPR scenario looked a little less like a thought experiment and more like one you might actually do.

Let's say that the particles are emitted in opposite directions. Once they have travelled for some time, we measure the respective property (polarization, spin) of one of them. We don't know what we'll measure until we measure it – but once we know the outcome, we can be sure also that the other particle has the opposite value.

This doesn't seem at first like such a big deal. Think of a pair of gloves: one left-handed, the other right-handed. If we were to post one at random to Alice in Aberdeen and the other to Bob in Beijing (I'd happily call him Bai or Bo, but conventions are conventions), then the moment Alice opened the parcel and found the left glove (say), she'd know that Bob's glove is right-handed. This is trivial, because the gloves had that handedness all the time they were in transit – it's just that Alice and Bob didn't know which was which until one of them looked.

But quantum particles are different – at least, that's what Bohr was insisting. In the Copenhagen Interpretation, spins and photon polarizations are undefined until a measurement is made on them. Until that point they don't *have* any particular value. Yet still quantum entanglement imposes the correlation between the values for the two particles in the EPR experiment. So now if Alice measures one photon (say) and finds it has a vertical polarization, she has *elicited* that polarization by making the measurement. Yet Bob's photon must then have a horizontal polarization, and this too has seemingly been imposed by Alice's measurement. And there seems no avoiding the conclusion that it must happen the instant Alice makes her measurement.

It takes light about a fortieth of a second to travel from Aberdeen to Beijing. It would be easy, with today's optical technology and accurate timekeeping, to arrange matters so that Bob measures the polarization of his photon after

Alice measures hers but before light could have travelled from Aberdeen to Beijing. Yet still – according to quantum mechanics – Alice and Bob will observe a correlation between the orientations of their photons. It's as if the effect of Alice's measurement is communicated to Bob's photon *faster than light*.

While the details differed, this 'instantaneous communication' was what the original EPR paper identified as a prediction of quantum mechanics. But that's impossible, the authors said, because Einstein's theory of special relativity forbids any signal to travel faster than light. If Bohr was right that quantum objects don't have properties at all until they are measured, the EPR experiment contains an impossible effect: what Einstein called 'spooky action at a distance'. This was the EPR paradox.

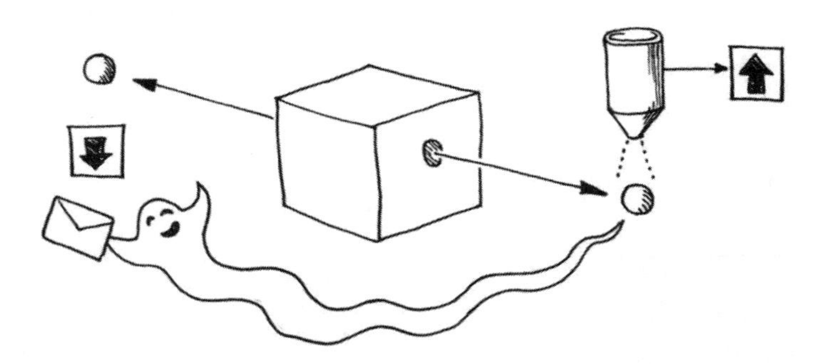

David Bohm's version of the EPR experiment, in which a measurement of the spin of one of a pair of entangled particles seems, according to quantum mechanics, instantaneously to influence the spin of the other – as though some 'spooky' message is sent between them.

What, though, if the photon polarizations were already determined from the outset by hidden variables, only to become manifest when the measurements were made? Then there's no problem: we're back with the gloves.

The trouble is that there *are* no hidden variables in quantum mechanics that 'secretly' assign definite values to variables even though they appear to acquire them randomly through the act of measurement. Well then, said EPR, quantum mechanics must be missing something. As Einstein wrote to Max Born in 1948:

•

After some consideration, Bohr realised that Einstein had identified a serious problem. It was by thinking of a way out that he arrived at the first clear formulation of what, in his view, measurement is. It makes no sense, he said, to speak of mechanisms going on 'behind' the measurement, with this particle 'communicating' with that one. Those are precisely the kinds of 'underlying microscopic phenomena' that quantum mechanics commands us to forsake. The measurement *is* the phenomenon, and quantum mechanics predicts the outcome reliably.

This, however, wasn't really much more than a restatement of his previous position – and it seemed a touch evasive. What EPR were proposing seemed quite clear. There is Alice's particle, and there is Bob's particle, and *observing one seems instantly to affect the state of the other.* Alice measures only her particle; why should we be forced to include Bob's particle as part of that phenomenon, when it is all the way over in Beijing – or conceivably, if we delay measurement for long enough, even on another planet or in another solar system?

But if the EPR paradox was discomfiting, wasn't it verging on metaphysics anyway? Even if you could do the experiment, what would you learn? If Alice and Bob found that the polarizations of their entangled photons showed the expected correlation, that alone wouldn't tell you if it was because of spooky action at a distance in quantum mechanics or because of Einstein's hidden variables that fixed the polarizations all along. How could one discriminate between these two possibilities?

In 1964 an Irish physicist named John Bell showed how it could be done. Bell had a day job as a particle physicist at CERN in Geneva. Thinking about fundamental problems in quantum mechanics was just a sideline: as he famously quipped, 'I am a quantum engineer, but on Sundays I have principles.' Yet Bell thought about those principles more deeply than almost anyone besides Bohr.

He reformulated the EPR experiment in a way that suggested how it could be conducted using available technology, and which should produce different results if indeed quantum mechanics alone described the situation or if something extra like hidden variables were needed. Bell, like Einstein, was rather sceptical that quantum mechanics was adequate.

Bell's proposal involved making repeated measurements on pairs of entangled particles. If a certain combination of these experimental results fell outside a particular numerical range, hidden variables were impossible and quantum mechanics needed no such modification.

Bell's experiment amounts to enumerating the difference between how strongly correlated two particles can be if entanglement involves hidden variables and if it is simply described by quantum mechanics as it stands (however peculiar that seems). At first glance these two ideas seem to be describing the same thing: how a measurement on one particle can turn out to be correlated with a measurement on the other. Bell's genius was to find a way of teasing out a *difference* in the predictions of the two models, which can then be measured. He showed that purely quantum-mechanical correlations can be stronger than those permitted by hidden variables.

Hidden variables, remember, fix matters so that the particles have definite states all along, although we don't know which is which until we measure them. It's not immediately obvious how one could get a stronger correlation than that: if one glove is left-handed, the other must be right-handed. But by appearing to allow particles to communicate so that, you might say, they can exchange notes instantly and decide on their state accordingly, quantum mechanics *does* permit an even stronger degree of correlation – which should show up in the measurement statistics.

Bell framed his analysis in terms of particles with correlated spins, such as electrons. Alice and Bob may measure these spins using magnets in a Stern–Gerlach-style experiment (page 135). Such a measurement, we saw, can only give one of two values (*up* or *down*), regardless of the relative orientation of the magnets and the spins.

Remember, though, that to see behind the randomness of quantum measurements we need to make many of them for identical experiments, and then take the average. Only that way did we see previously that when we measure the horizontal component of a vertically aligned spin, it has the average value of zero – even though each individual measurement gives the non-zero *up* or *down* at random.

Bell's experiment is rather like that one, but conducted on two entangled particles with anti-correlated spins, meaning that if one is *up*, the other is *down*. There are four possible outcomes for any given pair of particles: the spins measured by Alice and Bob are {*up, up*}, {*down, down*}, {*up, down*} or {*down, up*}. In the first two cases there's complete correlation between the spins: we can assign the correlation a value of +1. In the latter two cases there is complete *anti*-correlation (–1). And *these are the only possible outcomes* for each measurement.

The key is that a hidden-variables picture and a quantum-mechanical picture give different predictions of how the strength of correlation depends on the angle between the orientation of Bob's magnets and Alice's. If they measure the two spins using magnets in the same orientation, they find perfect anti-correlation (–1) each time (because that's the way the particles are produced): if one spin is *up*, the other is *down*. So the average over many measurements is also –1. If one of the magnets is rotated by 180° relative to the other, now both Alice and Bob always measure the same spin orientation: the correlation is always +1.

If Alice's magnets are at a right angle relative to Bob's, however, there no longer seems to be any correlation at all – the relationship between the spins is zero on average (although it must be +1 or –1 in any individual experimental run), just as the x component of a single spin aligned in the z direction averages to zero. And as we saw, this is ultimately a consequence of the Uncertainty Principle: in effect, you could say that in this configuration Alice and Bob can't find out about the spin of one particle in the pair without relinquishing all knowledge of the other. If Alice measures her particle, she can no longer use that measurement to deduce anything about Bob's, and vice versa.

So then, we know what the average correlations are for magnets aligned at relative angles of 0° (–1), 90° (0), 180° (+1) and 270° (0). What about the angles in between? For a hidden-variables model, we can show that the degree of correlation is simply proportional to the angle. But for quantum mechanics, it turns out that the degree of correlation is predicted to depend on the cosine of this angle. If you have forgotten your trigonometry, all this means is that the relationship is a curve rather than a straight line. So the predictions are different.

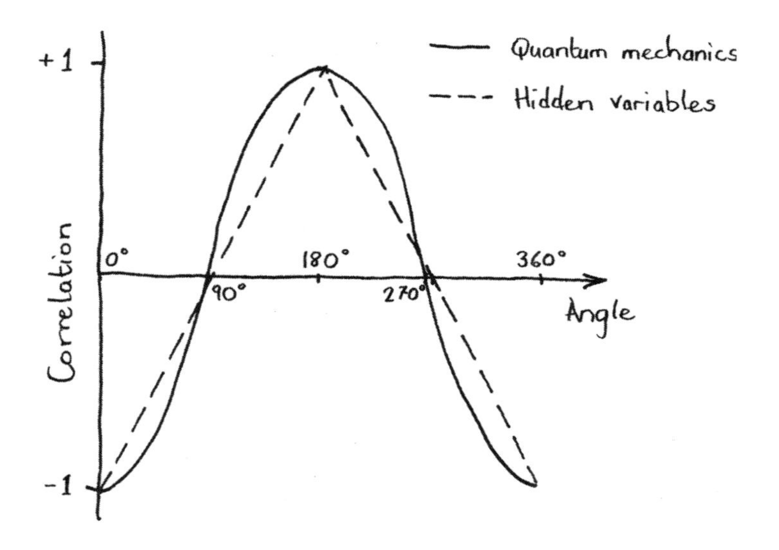

How the average correlations between Alice and Bob's spin measurements depend on the angle between their magnets in the Bell experiment. The predictions of quantum mechanics differ from those of a hidden-variables model for angles in between right-angle orientations.

This is actually a somewhat simplified version of Bell's scenario. He proposed a situation in which Alice and Bob can switch between two different measurement angles for each run, and then they combine the measured average correlations of the four possible measurement geometries in such a way that, for a hidden-variables description, their sum has to lie within the range +2 to –2 for all values of the angle.* But when you use quantum mechanics to predict the outcome, you find that the average value of the summed correlations *can* lie outside the range +2 to –2. The details don't matter; the principle is the same.

* This limit is, more strictly, that for a particular implementation of Bell's experiment, described in 1969 by John Clauser, Michael Horne, Abner Shimony and Richard Holt.

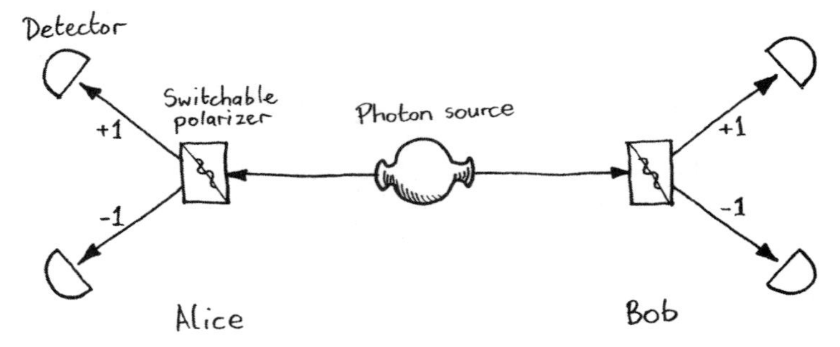

How to realize Bell's experiment for investigating the 'EPR paradox' using laser photons. Here Alice and Bob perform simultaneous measurements on the correlations (either +1 or –1) between the polarization of two entangled photons. The experiment involves measuring how this correlation changes as the relative orientation of the direction in which polarization is measured for the two photons is changed. If each photon's polarization is fixed from the outset by hidden variables, a particular sum of the four possible measurement outcomes for each experimental run must remain within the bounds +2 to –2. But if quantum mechanics alone governs the outcome, the sum can fall outside of these bounds.

However, it's when we consider this version of Bell's experiment – with four possible measurements on each run and a bound on the statistics that quantum mechanics may exceed – that we see what's so very peculiar about it.

You see, on the one hand this is just a classic example of how science works. You have two rival explanations of a phenomenon, and so you come up with an experiment in which the two theories give different answers – and see which is correct.

But here's the problem. The enumeration of outcomes is based on the fact that, in each individual experimental run, the two entangled spins can *only* yield a correlation of +1 or –1. Nothing else is possible. If that is so, Bell's combination of outcomes seems guaranteed to be restricted to the values between –2 and +2 – not by some physical law, but by the simple rules of arithmetic. That's how it is constructed.

In other words, the quantum prediction appears to violate basic arithmetic. How can it do that?

The catch is that, to calculate the bounds on Bell's sum, we're assuming that the spins do have to take the values *up* or *down* (that's to say, for electrons the spins are +½ and –½). Well, so what? Haven't I said that this must be true anyway, because they are quantized?

Not quite. I said that whenever we measure an electron spin, it must have the value ±½. This seems to mean that indeed the only possible values for the correlation between them are +1 and –1. But in any experimental run, there are four possibilities for this correlation. Yet Alice and Bob only ever measure two of them: each chooses to set the angle of the magnets to either this value or that one. If we'd measured either of the other two options instead, we can be certain that we'd get a value of ±1 too. *But we didn't measure them!*

Again, so what? If we know that we can only ever get a particular result from a measurement, what difference does it make if we don't actually do the measurement? I'm not, after all, saying that perhaps every now and then the measurement will sneakily yield a value that is *not* equal to ±1. It won't, as far as we can know.

No, the problem is that we're assuming we can say something meaningful about a quantity that we don't measure. But in the Copenhagen Interpretation, we can only make meaningful statements about things that we *do* measure. As Asher Peres has put it, 'Unperformed experiments have no results.' It is the inability to speak meaningfully about a quantity we don't measure that allows quantum mechanics to violate Bell's bounds.

There can be no clearer demonstration that Bohr's refusal to contemplate any meaning for things not observed wasn't just stubborn pedantry. If he's right, there are *measurable consequences*. This wouldn't mean that the Copenhagen Interpretation is correct, but it would mean that Einstein's hidden variables – any attempt, in fact, to fix everything in principle in a quantum system before it is observed – won't wash.

Thanks to Bell, it became possible to put the EPR experiment to the test and find out who was right. This has now been done many times. Every single time, the observed correlation statistics have turned out to match what quantum mechanics predicts, and to rule out Einstein's hidden variables. Yes, it seems Einstein was 'wrong' to consider that his thought experiment highlighted a fatal flaw for quantum theory.

But what, then, of the 'EPR paradox' and its spooky action at a distance?

There is no ‘spooky

action at a distance'

The modern quantum renaissance can be fairly said to have started in the 1960s with John Bell's work on entanglement. But just as it was when quantum mechanics itself was launched in the early 1900s by Planck and Einstein, it took the world some time to catch up.

For this renaissance Einstein again deserves some of the credit, albeit rather indirectly. In 1917 he showed that the quantum theory of light emission from energetically excited atoms predicts that if there's a bunch of them, all the photons can be released at once in a kind of avalanche, with their waves synchronized. In 1959 this effect was christened light-amplified stimulated emission of radiation – a cumbersome term that was condensed into the euphonious acronym LASER. In the early 1960s researchers figured out how to achieve laser action experimentally, first with microwaves and then with visible light. The laser, which permits exquisite control over photons, has become the central device for making quantum thought experiments practical. It, more than anything else, has allowed us to start exploring, and not just speculating about, the fundamentals of quantum mechanics.

By the 1970s, lasers offered a means to carry out Bell's test of quantum entanglement. The experiments were extremely challenging. The first to attempt them were two physicists, John Clauser and Stuart Freedman, at the University of California at Berkeley. They used lasers to coax entangled pairs of photons with correlated polarization states out of excited calcium atoms, and set about measuring the EPR correlations between them using the 'four state' set-up I showed earlier.

Clauser and Freedman found correlations stronger than those permitted by hidden variables in Bell's theorem. But their results weren't entirely clear-cut; for one thing, there weren't enough experimental runs for the statistics to be totally persuasive. A more definitive demonstration that entanglement fits with quantum mechanics, but not with hidden variables, was conducted in 1982 by Alain Aspect and co-workers at the University of Paris in Orsay. They too used laser and fibre-optic technologies to generate and manipulate entangled photons.

Recall that the Bell test involves enumerating the correlations between particles at different measurement angles. Aspect and colleagues were able to address a loophole in Bell's argument: the possibility that (by some unknown mechanism) the filters used to make the measurements of photon polarization were somehow interacting and influencing one another, artificially boosting the apparent quantum correlations. The French team could change the filter orientations rapidly during the time it took for the photons to leave their source and arrive at these detectors. Once a photon is heading towards the filter, no influence from the other filter can outrun it and switch the filter setting in advance.

It seemed, then, that quantum mechanics was correct. In which case, what is entanglement telling us? The mystery of the EPR experiment, says David Mermin, is that 'it presents us with a set of correlations for which there is simply no explanation'. All quantum mechanics gives us is a prescription for them. Can that really be enough?

•

First, we'd better confront that 'paradox'. If indeed the properties of particles are indeterminate until one is measured, it does look as if there is instantaneous communication between them in an EPR experiment. The unobserved particle seems to 'know' at once which spin or polarization the measurement on the other particle has produced, and to then adopt the opposite orientation. Contrary to what Einstein thought, however, that is not really 'action', it is not 'spooky', and it doesn't exactly involve 'distance'. Neither does it violate special relativity.

What relativity says is that events at one place may not exert a *causal* influence on events at another place faster than the time it takes for light to pass between them. By *causal* I mean that Alice does something and it determines what Bob sees. Only if this is so can Alice use a correlation between their observations to communicate with Bob.

Consider now Bohm's version of the EPR experiment for two particles with correlated spins. Alice makes choices in how she makes her observations – the angle of her magnets in a Stern–Gerlach measurement of spin, say – and these show up as correlations with what Bob measures. But they can *only* deduce that this is so by comparing their measurements: that is, by exchanging information by normal means, which can't happen faster than the speed of light. Bob can't discover what Alice has measured any faster than this.

So while it is possible for Alice and Bob retrospectively to *infer* something that looks like instant – and yes, spooky – action at a distance, they can't use this phantasmal connection to send any information faster than light. Let's say that

Alice's and Bob's particles are anti-correlated (their orientations must be opposed), and Alice is going to try to exploit that, using her magnet orientations, to send an instantaneous message to Bob. If Bob measures an *up* spin, he doesn't know if it's because Alice's particle was *down* and her magnets were aligned with his, or because her particle was *up* too and the magnets were anti-aligned, or because her magnets were at right angles and so there's no correlation between her particle and his. His subsequent measurements give a sequence of *ups* and *downs*, but he can't deduce from them anything that Alice is doing with her magnets.

Wait! Isn't it still the case that Alice is causing Bob's results by her choices, but he just can't understand what she's saying? No it isn't. Alice hasn't caused Bob's spin to be *up* on any occasion, because she can't even fix her own spin. It could be *up* or *down*, at random. Nothing Alice does determines what Bob sees: there's no 'action at a distance', and special relativity is intact.

But still the correlation appears when they compare their measurements! Where did that come from? As Mermin says, there is 'no explanation' – or rather, we might say, it came from 'quantumness', about which we can't construct a narrative.

Although this argument is scientifically sound, one can't avoid the feeling that we have violated relativity in spirit while concocting a logical argument to deny it. Even if relativity emerges (by the skin of its teeth) unscathed, there's still something uncanny about quantum entanglement – because it undermines our preconceptions about the here and now, the here and there. It messes with time and space.

•

It took many years to figure out what was wrong with Einstein's reasoning about the EPR 'paradox'. That's because, as is so often the case with quantum mechanics, the problems lie buried beneath what looks like plain common sense.

Einstein and his colleagues made the perfectly reasonable assumption of *locality*: that the properties of a particle are localized on that particle, and what happens *here* can't affect what happens *there* without some way of transmitting the effects across the intervening space. It seems so self-evident that it hardly appears to be an assumption at all.

But this locality is just what quantum entanglement undermines – which is why 'spooky action at a distance' is precisely the wrong way to look at it. We can't regard particle A and particle B in the EPR experiment as separate entities, even though they are separated in space. As far as quantum mechanics is concerned, entanglement makes them both parts of a single object. Or to put it another way, the spin of particle A is not located solely on A in the way that the redness of a cricket ball is located on the cricket ball. In quantum mechanics, properties can be *non-local*. Only if we accept Einstein's assumption of locality do we need to tell the story in terms of a measurement on particle A 'influencing' the spin of particle B. Quantum non-locality is the *alternative* to that view.

What in fact we're dealing with here is another kind of quantum superposition. We've seen that superposition refers to a situation in which a measurement on a quantum object could produce two or more possible outcomes, but we don't know which it will be, only their relative probabilities. Entanglement is that same idea applied to two or more particles: a superposition of the state in which particle A has spin *up* and B spin *down*, say, and the state with the opposite configuration. Although the particles are separated, they must be described by a single wavefunction. We can't untangle that wavefunction into some combination of two single-particle wavefunctions.

Quantum mechanics is able to embrace such a notion without batting an eyelid; we can simply write down the math. The problem is in visualizing what it means.

•

Because quantum non-locality is so counter-intuitive, scientists have gone to extraordinary lengths to verify it. Might we not be overlooking something else that merely creates an illusion of non-locality?

In testing one such loophole, Aspect's experiment proved to be just the opening act of a series of investigations that is still ongoing. A possible influence – fast but necessarily slower than light – between the detectors, which Aspect and collaborators considered and excluded, is now called the locality or communication loophole. What kind of influence could do that anyway, you might ask? Well, who knows? The quantum world is full of surprises. You can't just say it's impossible until you've looked.

We can rule out this loophole with even more confidence than Aspect did by increasing the distance between detectors (including the polarizing filters for photons) until no signal slower than light could possibly have time to travel between the two before the entire measurements are completed. Researchers at the University of Innsbruck in Austria achieved that in 1998 by placing the detectors 400 metres apart, providing enough time for the wizardry of optical technology to complete its job before any communication could pass between the measurement sites. They found no change in the outcome of the experiment.

Then there's the 'freedom of choice' loophole, in which some local property of the particles themselves, 'programmed' when they are placed in an entangled state, influences the setting of the detectors when the measurement is made. This was ruled out in 2010 in an experiment that closed the locality loophole at the same time, by making sure that the detectors were distant not only from one another but also from the photon source: the source and one of the detectors were located on separate islands in the Canaries. These experiments, incidentally, show us something else important about quantum effects like entanglement: they persist over macroscopically large distances. This is one reason why it is not quite accurate to call quantum mechanics a theory only of the 'very small'. It will work between me and you, wherever you are.

The 'fair sampling' or detection loophole, meanwhile, allows the possibility that some local property of the particles biases their detection so that we don't get a truly random sampling. In any Bell experiment the detection is imperfect: only some of the particles are measured. To give a reliable result, this subset has to be representative of the whole. Ruling it out demands a high detection efficiency, so that you can be more confident you're seeing the whole picture.

Admittedly, it would be remarkably bad luck if inefficient detection of particles turned out to give results perfectly mimicking what quantum mechanics predicts, while a better detection method would unmask a departure from those predictions.

All the same, who knows? So researchers led by Anton Zeilinger at the University of Vienna checked in 2013. They had a more efficient way of detecting the particles (photons), which captured about 75% of them. This is still less than the threshold above which it's possible to be absolutely confident that a violation of Bell's inequality has genuinely been observed in the kind of EPR experiment described earlier. But Zeilinger and colleagues also used a variant of Bell's theorem that cleverly builds into the equations the possible effects of those particles that remain undetected. In this case the detection threshold above which quantum mechanics can be considered to be validated is lowered to just 67%. So the experiment had the discriminating power needed to rule out the detection loophole. Which is just what it did.

Are there any loopholes left? It's getting harder and harder to dream up anything plausible. Ah, but what if different loopholes are operating in different experiments? Now you're really grasping at straws! Still, I suppose we'd better check. And so the game now is to close several loopholes simultaneously. In 2015 a team at the Technical University of Delft in the Netherlands, led by Ronald Hanson, excluded at the same time both the communication loophole and the detection loophole in an ingenious tour de force. It got around the detection loophole by making measurements on entangled electrons, which can be more reliably detected than photons of light. And it closed the communication loophole by linking the electrons' entanglement to that between photons, which can be transmitted over long distances (in this case 1.3 kilometres) along optical fibres. Teams in Austria and Boulder, Colorado have also reported experiments that close these two loopholes at the same time.

The Dutch result was greeted with the usual headlines announcing that 'Einstein was wrong' because 'spooky action at a distance is real'. Now you know better.

•

It has been proposed, albeit in a highly speculative theoretical scenario, that the interdependence across space that manifests as quantum entanglement is what stitches together the very fabric of space and time, creating the web that allows us to speak of one part of spacetime in relation to another. Spacetime is the four-dimensional fabric described by Einstein's theory of general relativity, in which it is revealed to have a particular shape. It's this shape that defines the force of gravity: mass makes spacetime curve, and the resulting motions of objects in that curved arena make manifest the force of gravity. In other words, entanglement could be the key to the long-standing mystery of how to reconcile quantum mechanics with the theory of gravitation supplied by general relativity.

In some simple models of a quantum universe, a phenomenon that looks like gravity emerges spontaneously from the mere existence of entanglement. Physicist Juan Maldacena has shown that a model of an entangled quantum universe with only two dimensions of space and lacking any force of gravity at all mimics the same kind of physics seen in a three-dimensional model of an 'empty' universe filled with the kind of spacetime fabric necessary for a general-relativistic description of gravity. That's a mouthful, but what it amounts to is that taking away entanglement in the 2D model is equivalent to unweaving spacetime in the 3D case. Or, you might say, spacetime and gravity in the 3D universe look like a projection of the entanglement existing within its 2D boundary surface. Without that entanglement all around the edge, spacetime unravels and the 3D universe splits up.

This theory is far too simplified to describe what goes on in our own universe, and so these ideas are still very tentative. But many researchers suspect that this deep connection between entanglement and spacetime is telling us something about how quantum mechanics and general relativity are related: about what needs to change in our view of spacetime if quantum theory is to be made consistent with general relativity. David Bohm anticipated this idea decades ago when he suggested that quantum theory hints at an order somehow connected to what we have regarded as spacetime, yet richer. Some researchers now suspect that spacetime is actually *made* from the interconnections created by quantum entanglement. Others think there is more to it than that.

Regardless of how such ideas will fare, there's a developing suspicion among physicists that a quantum theory of gravity isn't simply going to come from clever math, but requires us to think in a different manner about both quantum mechanics and general relativity. Spacetime is really just a fabric we posit to describe how one thing affects another, and to express the limitations on such interactions. It's an emergent property of *causal* relationships. And as we've now seen, quantum mechanics forces us to revise our preconceptions about causation. Non-locality, entanglement and superposition appear not just to allow objects to become interconnected in a way that pays no heed to spatial separation but also seems to do odd things to time, such as producing an illusion (or perhaps more than that?) of backwards causation, or allowing superpositions of the causal ordering of two events (so that which came first is indeterminate; see page 280).

It could be that the *causal structure* of the universe is a more fundamental concept than both of these theories. We'll see later why this causal structure could be a good place to start in rebuilding quantum mechanics from the ground up in a way that makes its axioms more physically meaningful and less abstract and mathematical.

•

In 1967, three years after John Bell proposed his theorem introducing the concept of quantum non-locality, a related counter-intuitive aspect of quantum mechanics was identified by the mathematicians Simon Kochen and Ernst Specker. Their work, while equally fundamental, has received far less attention until relatively recently. (John Bell had in fact already understood the same point as Kochen and Specker, but his own proof, formulated in 1966, wasn't published until after theirs.)

Kochen and Specker pointed out that the outcomes of quantum measurements may depend on their context. This means something subtly different from the fact that different types of experiment on what seems like the same system (a double-slit experiment with or without a 'path detector', say) give different results. It means that if you look at a quantum object through different windows, you see different things.

If you want to count the number of black and white balls in a jar, it doesn't matter if you count the black ones or the white ones first, or if you tot them up in rows of five, or if you separate the two colours into piles and weigh them. You'll get the same answer. But in quantum mechanics, *even when you ask the same question* ('How many black and white balls are there?'), the answer you get may depend on how the measurement is done.

We already saw that making successive measurements on quantum objects in different orders – first *this*, or first *that* – can give different outcomes. And this is a consequence of the fact that the respective mathematical operations conducted on a quantum wavefunction in order to extract from it values of observable properties do not commute (page 151).

The Kochen–Specker theorem stipulates what follows from this dependence on context. In effect it is another corollary of the way that, in quantum mechanics, what we choose *not* to measure may leave an imprint on what we *do* measure. It explores the consequences of what lies outside the particular window through which we elect to view the quantum system.

Specker had a story about that. An Assyrian seer, unwilling to let his young daughter be wed to suitors who he considered unworthy, set them a task. He presented them with three closed boxes in a row, each of which might or might not have a gem inside it. Any prediction will identify at least two boxes that are guessed to be both empty, *or* two that are both filled. (Think about it for a moment and you'll see that this must be so.) The suitors were asked to open those two, and if they were right then they would have the daughter's hand. But they were never right! One of the opened boxes was always empty while the other contained a gem. How could that be? Surely chance alone should guarentee that someone would guess right some time?

Finally the daughter, who was getting impatient to be wed, intervened by opening up the boxes for the rather dishy son of a prophet. But she didn't open the two he'd predicted to be both full or both empty. Instead she opened one he'd guessed was filled, and one he'd guessed was empty. And both guesses proved correct. After some weak objection, the seer accepted that the suitor had made two correct guesses, and the daughter was married.

The seer had foiled previous suitors because his boxes were quantum boxes, and he had entangled them to create correlations such that, if one opened box was filled, the other would be empty and vice versa. That made it impossible to satisfy the seer's challenge and show you'd guessed right. But the daughter was able to reveal a correct guess by interrogating the *same* system with a *different* set of measurements. And that is what quantum contextuality is like.*

Like Bell's theorem, the Kochen–Specker theorem says something about what hidden variables – hypothetical concealed factors that fix the properties of quantum objects irrespective of whether they are measured or not – would have to be like in order to deliver experimental outcomes that look just like those quantum mechanics predicts. Hidden variables, remember, are *local*: they apply specifically to this or that object, just as the properties of

* It turns out that it is not possible to implement exactly this scheme of the seer's using quantum mechanics. But you can arrange for something analogous. We'll come back to quantum boxes later.

macroscopic objects do. Bell's theorem provided the theoretical tools for assessing if such localized hidden variables are feasible, and the experiments insist that they are not.

Kochen and Specker posed an even stronger problem for hidden variables. They showed that you can't mimic the predictions of quantum mechanics (such as correlations between the properties of two particles) with hidden variables that relate only to the quantum system you're studying. You'd also have to include some hidden variables related to the *apparatus* used to study it. In other words, you can never say 'this system has such and such properties', but only that it has those properties in a particular experimental context. Alter that context and you alter the overall hidden-variables description.

So you can never use hidden variables to say 'what is real' about an object in every circumstance. You can't say, to make a macroscopic analogy, 'the ball is red', but only 'the ball is red when looked at through the round window'. It *really is* red under these conditions (to the extent that we can ever say how something really is). But equally, it *really is* green through the square window. Yes, but what colour is it *really*? According to Kochen and Specker, you can't improve on the answer I've just given.* To put it another

* I've chosen this example of colour precisely to give a flavour of what this 'contextuality' is all about – for of course in reality a macroscopic ball too can only be assigned a colour if you specify the kind of light illuminating it, and can seem to change colour when illuminated with different kinds of light. This is a subjective effect of vision, not anything truly connected to quantum contextuality: if we're more careful about defining what we mean by 'colour', we can say something about what it implies for the ball regardless of the context. But maybe this crude analogy can offer the mind an intuitive foothold.

way: for all the conceivable true-or-false propositions we might make about a quantum object – it is red, it is moving at 10 mph, it is spinning once a second – we cannot simultaneously give definite yes/no answers. Not everything is knowable – because not everything is *be*able – at once.

For reasons that are not easy to fathom, experimental studies of quantum contextuality lagged behind those of quantum non-locality by two or three decades, and the first experiments that clearly confirmed the Kochen–Specker theorem were performed only in 2011.

It's long been suspected that quantum non-locality and contextuality are somehow related. Dagomir Kaszlikowski of the National University of Singapore has suggested that indeed they are ultimately expressions of the same thing: different facets of a more fundamental 'quantum essence', for want of a better term. This essence, whatever it is, defies any *local realist* description of the quantum world: one in which objects have specific, well-defined features that are intrinsic to the object itself. You simply can't say – as we're accustomed to do in the macroscopic world – 'this thing here is like *so*, regardless of anything else'.

Kaszlikowski and his colleagues showed that non-locality and contextuality seem in fact to be mutually exclusive: a system can exhibit one feature or the other, but never both at the same time. That's to say, 'quantumness' can enable a system to exceed the hidden-variables correlations in a Bell-type experiment, or it can enable the system to show a stronger dependence on context of measurement than a hidden-variables model can accommodate. But it can't do both at the same time. Kaszlikowski and colleagues call this behaviour monogamy.

So what is this 'essence' that can manifest itself as either this or that kind of counter-intuitive behaviour? We don't know. But simply arriving at that question is an advance in our understanding: as ever in science, a big part of the art is finding the right way to express a problem.

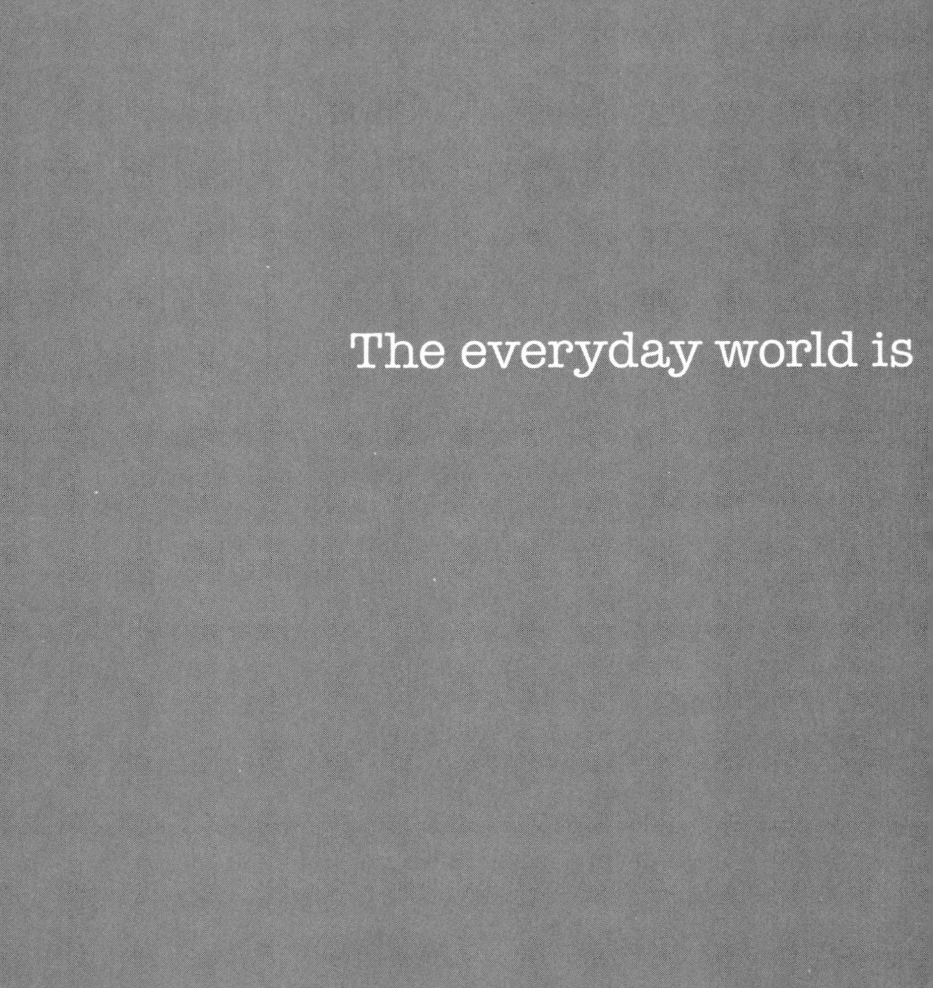
The everyday world is

what quantum becomes at human scales

I suggested earlier that the one thing 'everyone knows' about quantum mechanics is that the quantum world is fuzzy and uncertain. Actually there's another thing too. 'Everyone' has heard of Schrödinger's cat.

How seriously should we take Schrödinger's cat?

Doubtless that's why there are jokes about this idea too. Schrödinger is driving along when he is pulled over by a policeman. The officer looks over the car and asks Schrödinger if he has anything in the boot.

'A cat,' Schrödinger replies.

The policeman opens the boot and yells 'Hey! This cat is dead!'

Schrödinger replies angrily, 'Well, he is now.'

Don't worry, I'm not going to pedantically pick this one apart too. As physics jokes go, it's not so bad. At the least, it illustrates what a good job Schrödinger did in finding an image catchy enough to become a cultural meme.

Schrödinger's aim was to demonstrate the paradox created if we try to divide the world into classical and quantum parts. What happens when the two can't be so neatly separated?

But there are many uses for a cat, and Schrödinger wasn't simply illustrating the problem of having one rule for the big and another for the very small. He wanted to demonstrate the apparent *logical* absurdity of quantum mechanics, in which mutually contradictory or exclusive circumstances – such as live and dead – coexist.

One might say that Schrödinger's metaphor was *too* successful. The cat is still hauled out today as if to imply that we're as puzzled as ever by the mere fact that the quantum world at small scales turns into the world of classical physics at human scales. The fact is, however, that this so-called 'quantum–classical transition' is now largely understood. Things have moved on, and we can state much more precisely than could Schrödinger and his contemporaries why and how quantum becomes classical. The answer is both elegant and rather astonishing.

For quantum physics is not replaced by another sort of physics at large scales. It actually *gives rise* to classical physics. Our everyday, commonsense reality is, in this view, simply what quantum mechanics looks like when you're six feet tall. You might say that it is quantum all the way up.

The question, then, is not why the quantum world is 'weird', but why ours doesn't look like that too.

•

In Schrödinger's day it seemed fanciful to imagine that we could ever have a direct window onto the borderland between the microscopic and the macroscopic. It seemed even less likely that we could assert any control over that liminal region. So it was acceptable to pretend that this boundary was absolute, even if its location was somewhat hazy and open to debate. The quantum–classical transition is then like an ocean crossing between two continents: drawing a border somewhere in the open sea is an arbitrary exercise, but the continents are undeniably distinct. The land of the quantum, said Schrödinger, is random and unpredictable, yet the classical realm is orderly and deterministic because it depends only on statistical regularities among that atomic-scale chaos.*

Yet we are no longer compelled to navigate from quantum to classical with our eyes closed, but can undertake the journey while taking in every nuance of the shifting view. Advances in technology have made it possible to perform experiments that probe the quantum–classical interface from both directions, moving

* An exception, said Schrödinger, is found in the living world, where somehow order is maintained at the molecular scale: life is not, he said, 'order from disorder', but 'order from order'. A single molecular event such as a chromosomal mutation, presumably governed by quantum rules, can produce a definite macroscopic effect. Schrödinger's musings about how this could be so led to his 1944 book *What Is Life?*, which was profoundly influential among the generation of biologists who elucidated the molecular principles of genes and heredity.

up in scale from the microscopic and down from the macroscopic. They meet at the 'middle scale' or so-called mesoscale, where we can now literally watch quantum become classical.

This experimental access doesn't just permit but *compels* an investigation of the quantum–classical transition. It's on this mesoscale, where distances are typically measured in nanometres (millionths of a millimetre) and atoms are counted in thousands and millions, that both nanotechnology and molecular biology operate. If we wish to intervene at this scale for practical purposes – to solve engineering and medical problems, say – then we must ask what kind of theory we should use: quantum or classical physics, or maybe a bit of both?

What has truly transformed our view of the quantum–classical transition, however, is a theoretical rather than a practical advance. Scientists have realized that they need to take account of an ingredient that the pioneers of quantum theory overlooked – even though it was literally all around them.

•

Erwin Schrödinger dreamed up his 'diabolical' (his word) thought experiment in 1935. It was intended as a challenge to Bohr's interpretation of quantum mechanics, towards which Schrödinger shared a great deal of Einstein's scepticism, and it arose from correspondence with Einstein after the publication of the EPR paper describing entanglement.

It was all very well for Bohr to impose a strict separation of quantum and classical, and to make observation the process by which they are distinguished – but what then if the quantum and the macroscopic are coupled without any observation taking place? Schrödinger was looking for what he called a 'ridiculous case': a *reductio ad absurdum*, not to be taken literally, in which we are confronted by a superposition of macroscopic states that seems not just bizarre (such as a large object being in 'two places at once') but logically incompatible. Einstein raised the prospect of a keg of gunpowder being in a superposition of exploded and unexploded states, and Schrödinger upped the ante with his cat.

The cat is placed in a box containing some mechanism that will kill it – Schrödinger imagined a vial of poison that is broken to release deadly fumes – if triggered by a specific outcome of an event governed by quantum mechanics, such as the decay of a radioactive atom. It's conceptually neater to think of the quantum trigger as an atom with a spin, perhaps coupled to a magnetic mechanism. If it's an *up* spin, say, the mechanism breaks the vial, but for a *down* spin that doesn't happen. Then we prepare the atom in a superposition of *up* and *down* spins, and close the box. We seem obliged now, he wrote, to describe the entire system with a wavefunction 'having in it the living and the dead cat (pardon the expression) mixed or smeared out in equal parts' – and it stays that way as long as we don't make a measurement of (look at) the cat. And while we can write about 'a live and dead cat', it's not clear that this notion has any logical meaning.

The scenario forces us to confront the puzzling indeterminacy that we might be more complacently liable to accept while it is hidden away at scales too small to see directly. As Schrödinger put it:

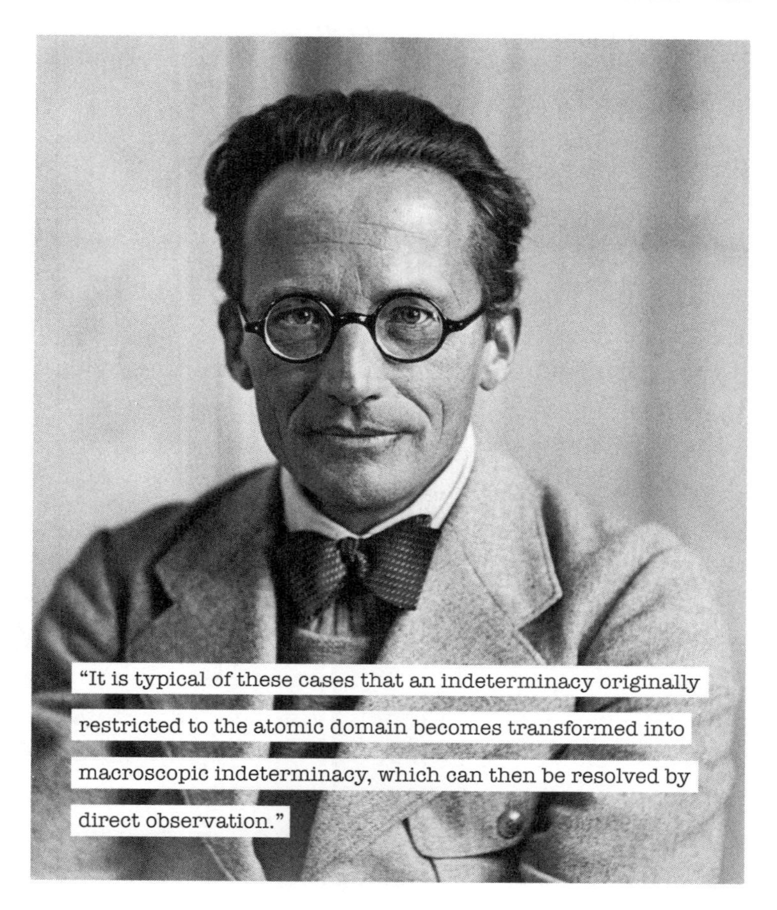

Schrödinger's thought experiment is eye-catching but overcomplicated. We can't easily say what qualifies as a live or dead cat – exactly when life becomes death isn't always straightforward even for doctors to decide. It would be clearer, perhaps, to pose the problem in terms of 'Schrödinger's dial' on which a needle points to two readings at once. We can't visualize that situation either – does it mean there would be a kind of ghostly half-image of the pointer needle? But of course the cat is what makes the experiment so infectiously arresting, so perplexing, so absurd.

Is there any way out of this ridiculous idea?

•

Schrödinger's cat forces us to rethink the question of what distinguishes quantum from classical behaviour. Why should we accept Bohr's insistence that they're fundamentally different things unless we can specify what that difference is?

We might then be inclined to point to features that classical objects like coffee cups have but which quantum objects don't necessarily have: well-defined positions and velocities, say, or characteristics that are localized on the object itself and not spread out mysteriously through space. Or we might say that the classical world is defined by certainties – either *this* or *that* – while the quantum world is (until a classical measurement impinges on it) no more than a tapestry of probabilities, with individual measurement outcomes determined by chance. At the root of the distinction, though, lies the fact that quantum objects have a wave nature – which is to say, the Schrödinger equation tells us that they should be described *as if* they were waves, albeit waves of a peculiar, abstract sort that are indicative only of probabilities.

It is this waviness that gives rise to distinctly quantum phenomena like interference, superposition and entanglement. These behaviours become possible when there is a well-defined relationship between the quantum 'waves': in effect, when they are in step. This co-ordination is called *coherence*.

It's a concept that comes from the science of ordinary waves. Here too, orderly wave interference (like that from double slits) happens only if there's coherence in the oscillations of the interfering waves. If there is not, there can be no systematic coincidence of peaks and troughs and no regular interference pattern, but just random, featureless variations in the resulting wave amplitude.

Likewise, if the quantum wavefunctions of two states are not coherent, they cannot interfere, nor can they maintain a superposition. A loss of coherence (*decoherence*) therefore destroys these fundamentally quantum properties, and the states behave more like distinct classical systems. Macroscopic, classical objects don't display quantum interference or exist in superpositions of states because their wavefunctions are not coherent.

Notice how I phrased that. It remains meaningful to think of these objects as *having* wavefunctions. They are, after all, made of quantum objects and so can be expressed as a combination of the corresponding wavefunctions. It's just that the wavefunctions of distinct states of macroscopic objects, such as a coffee cup being in *this* place and *that* place, are not coherent. Quantum coherence is essentially what permits 'quantumness'.

There is no reason (that we yet know of) why in principle objects cannot remain in coherent quantum states no matter how big they are – provided that no measurement is made on them. But it seems that measurement somehow does destroy quantum coherence, forcing us to speak of the wavefunction as having 'collapsed'. If we can understand *how* measurement unravels coherence then we would be able to bring measurement itself within the scope of quantum theory, rather than making it a boundary where the theory stops.

We might even be able to figure out what happens to Schrödinger's cat. (But I make no promises about that.)

•

Why it took so long for decoherence to appear as a core concept in quantum mechanics is not easy to say, given that the theoretical tools needed to understand it were around in Bohr's and Einstein's day. Perhaps this is simply another instance of how easy it is in this field to overlook the importance of what elsewhere we take for granted. For the crucial factor in understanding quantum decoherence is that ubiquitous entity present but largely ignored in all scientific studies: the surrounding environment.

Every real system in the universe sits somewhere, surrounded by other stuff and interacting with it. Schrödinger's cat might be placed inside a sealed box, but there must be air in there for the cat to have any chance of staying alive. And the cat is resting on a surface of some kind, exchanging heat with it. All of this sounds like a matter of detail that could, for the sake of argument, be neglected. If the environment is included at all in most scientific theories or analyses of experiments, it's often simply considered to be a source of a little random disturbance, which can, with sufficient care, be kept to a minimum.

But in quantum mechanics the environment has a central role in how things happen. It turns out to be precisely what conjures the illusion of classical physics out of the quantum soup.

It's often suggested that quantum states such as superpositions are delicate and fragile. Put them in a noisy environment (the story goes), and all that jiggling and shaking by the surroundings destroys these frail quantum states, collapsing wavefunctions and shattering superpositions. But this isn't quite right. Indeed, why *should* quantum states be fragile if, as I've suggested, quantum mechanics supplies the most fundamental description of the universe? What kind of laws are these, if they give up the ghost so easily?

The truth is that they don't. Quantum superpositions of states aren't fragile. On the contrary, they are highly contagious and apt to spread out rapidly. And *that* is what seems to destroy them.

If a quantum system in a superposed state interacts with another particle, the two become linked into a composite superposition. That, we saw earlier, is exactly what entanglement is: a superposed state of two particles, whose interaction has turned them into a single quantum entity. It's no different for a quantum particle off which, say, a photon of light bounces: the photon and the particle may then become entangled. Likewise if the particle bumps into an air molecule, the interaction places the two entities in an entangled state. This is, in fact, the *only thing that can happen* in such an interaction, according to quantum mechanics. You might say that, as a result, the quantumness – the coherence – spreads a little further.

In theory there is no end to this process. That entangled air molecule hits another, and the second molecule gets captured in an entangled state too. As time passes, the initial quantum system becomes more and more entangled with its environment. In effect, we then no longer have a well-defined quantum system embedded in an environment. Rather, system and environment have merged into a single superposition.

Quantum superpositions are not, then, really destroyed by the environment, but on the contrary infect the environment with their quantumness, turning the whole world steadily into one big quantum state. Quantum mechanics is powerless to stop it, because it contains in itself no prescription for shutting down the spread of entanglement as particles interact. Quantumness is really leaky.

This spreading is the very thing that destroys the *manifestation* of a superposition in the original quantum system. Because the superposition is now a shared property of the system and its environment – because the quantum system has lost its integrity and exists in a shared state with all the other particles – we can't any longer 'see' the superposition just by looking at the little part of it. We can't see the wood for the trees. What we understand to be decoherence is not actually a loss of superposition but a loss of our ability to detect it in the original system.

Only by looking closely at the states of all these entangled particles in the system and its surroundings can we deduce that they're in a coherent superposition. And how can we hope to do that – to monitor every reflected photon, every colliding air molecule? No – once the quantumness has leaked out into the environment, in general we'll never be able to concentrate the superposition back onto the original system. That's why quantum decoherence is, to all intents and purposes, a one-way process. The pieces of the puzzle have been scattered so widely that they are, *for all practical purposes*, lost – even though *in principle* they are still out there, and remain so indefinitely. That's what decoherence is: a loss, you might say, of *meaningful* coherence.

I slipped in one of those pedantic little qualifiers back there. I said that we'll never get coherence back *in general*. What I mean is that there's no law that absolutely forbids it; it's just impractical on the whole.* But if we are able to create a simple quantum system for which we can limit the rate of decoherence and keep careful track of this 'quantum spreading', we might be able to backtrack. This has been done: in very specialized situations scientists have observed the process of *recoherence*. For example, in 2015 physicists in Canada were able to recover the information lost through decoherence as two entangled photons moved through a crystal, and used it to reverse the entanglement of the photon pair with their environment – the process that induced decoherence – and recover the pristine state of the pair. It's the kind of exception that, in the proper sense, proves the rule.

•

Decoherence, then, is a gradual and real physical event that occurs at a particular rate. For some relatively simple systems, we can use quantum mechanics to calculate that

* Whether it is truly possible *even in principle* to undo decoherence (aside from some very special cases) is in fact disputed. If you do the math, you find that decoherence typically spreads out the superposition into the environment over more quantum states than there exist fundamental particles in the observable universe. Then you must face a philosophical question: can a problem be said to be strictly impossible solely because there is not enough information available in the universe to solve it?

rate: to work out how long it takes for decoherence to sabotage the possibility of observing, let's say, interference between the wavefunctions of a quantum object in two different positions. The further apart in space those two positions are, the more quickly the coherence between them will be destroyed by the environment – or rather we should strictly say, the faster it will become entangled with and leak away into the environment.

Take a microscopic dust grain floating in the air of my study, a hundredth of a millimetre across. How quickly do two positions of the grain decohere if their separation is (say) about the same size as the width of the grain itself, so that they don't overlap? Ignore photons for now – let's say the room is dark – and just think about the interactions between the grain and all the air molecules around it. Quantum calculations show that decoherence then takes about 10^{-31} seconds.

That's so short that we can almost say that decoherence is instantaneous. It happens in less than a millionth of the time it takes for a photon, travelling at the speed of light, to pass from one side of a single proton to the other.* So if you think you're going to see a quantum superposition of non-overlapping position states of a dust grain in my study, think again.

What if the same dust grain is in a complete vacuum, so that there are no collisions with air molecules? Still that doesn't suppress decoherence for long. A vacuum

* It's not obvious how to think about such a rapid process in physical terms, since a collision between the grain and an air molecule takes considerably longer. But we're talking here about the timescale created by many such collisions all going on more or less simultaneously.

at room temperature is being constantly criss-crossed by photons emitted from the warm walls of the vessel sustaining it. At everyday temperatures this thermal radiation is strongest at infrared frequencies. Interactions with these 'thermal' photons will induce decoherence of the dust grain in just 10^{-18} seconds, which is about the transit time of a photon across a gold atom.

Might we catch that superposition before it decoheres? It's perhaps within the realm of feasibility today, although it would be extremely difficult. But look, if thermal radiation is the problem then let's cool the environment down! Let's get rid of those thermal photons. We could conduct the experiment in space. Sure, there are some stray molecules even there, but let's assume we could get rid of them too. What's to induce decoherence then?

Even interstellar space, though, is not free of photons. They are humming about everywhere in the cosmos, in the form of the cosmic microwave background, the faint glimmer left over from the fury of the Big Bang itself. These photons alone – the remnants of creation – will decohere such a superposition of a dust grain in about one second.

The point is not that, *in extremis*, you can find a way to render observation of this 'mesoscopic' superposition feasible – if, that is, you can work out how to do it in space without actually disrupting the state in the measurement process itself. Yes, perhaps you could, and it would be an exciting prospect. But the point is that, for an object this size, you have to go to great extremes to avoid decoherence. For objects approaching the macroscopic scale under ordinary conditions, decoherence is to all practical purposes instant and inevitable.

And for microscopic objects? Well, then we really can avoid decoherence. That's the whole point – it's why we really can do experiments on atoms, subatomic particles and photons that reveal them to be (or to have previously been) in quantum superpositions. The figures tell the tale. For a large molecule (the size of a protein, say), decoherence happens within 10^{-19} seconds if it were floating in the air around us – but in a perfect vacuum at the same temperature it could stay coherent for more than a week.

Decoherence is what destroys the possibility of observing macroscopic superpositions – including Schrödinger's live/dead cat. And this has nothing to do with observation in the normal sense: we don't need a conscious mind to 'look' in order to 'collapse the wavefunction'. All we need is for the environment to disperse the quantum coherence. This happens with extraordinary efficiency – it's probably the most efficient process known to science. And it is very clear why size matters here: there is simply more interaction with the environment, and therefore faster decoherence, for larger objects.

In other words, what we previously called measurement can, at least in large part (not completely, as we'll see), be instead called decoherence. We obtain classical uniqueness from quantum multiplicity when decoherence has taken its toll.

Here is the answer to Einstein's question about the moon. Yes, it is there when no one observes it – because the environment is already, and without cease, 'measuring' it. All of the photons of sunlight that bounce off the moon are agents of decoherence, and more than adequate to fix its position in space and give it a sharp outline. The universe is always looking.

•

One possible reason why it took so long for decoherence to be identified as the mechanism for turning a quantum system 'classical' is that the early quantum theorists couldn't get past an intuition of *locality*: the idea that the properties of an object reside on that object. This is what entanglement undermines, and yet for many years after the EPR experiment had been proposed and debated there remained a presumption of neat separation between a quantum system and its environment, just as there is in classical physics. It wasn't until the 1970s that the foundations of decoherence theory were laid by the German physicist H. Dieter Zeh. Even Zeh's work was largely ignored until the 1980s, when the term 'decoherence' was coined. Interest in the 'decoherence program' was awakened by two prominent papers in 1981–2 written by Wojciech Zurek at Los Alamos National Laboratory in New Mexico, a former student of John Wheeler's.

The decoherence argument looks plausible in theory – but is it right? Can we actually see quantum effects leaking away into the environment? Serge Haroche, a French optical physicist, and his colleagues at the Ecole Normale Supérieure in Paris put the idea to the test in 1996. They studied bunches of photons held in a type of light trap called an optical cavity, in which the photons bounce around but cannot leave. The researchers passed a rubidium atom through the cavity in a superposition of two states. In one state, the atom interacted with the photons to cause a shift in their electromagnetic oscillations, whereas the other state caused no change. Then, when the experimenters sent a second atom through the cavity, it was affected by the state of the photons induced by the first atom. But that effect became weaker as the quantum state of the photon field decohered, so the signal produced by the second atom depended on the time delay relative to the passage of the first. In this way, Haroche and his colleagues could watch decoherence set in by altering the timings of the two atoms.

This procedure sounds complicated but amounts to creating a well-defined quantum superposition of photons at some moment in time, and then probing its decoherence at successive instants. It's very crudely like watching the decay of vibrations in a spring that is stretched and then released, as the spring's vibrational energy is dispersed. Other experiments that monitor the decay of coherence in superposed states have been conducted in very different systems, such as electronic components called superconducting quantum interference devices (SQUIDs – see page 264).

There's not much scope for controlling decoherence in these experiments: you get what you get. In 1999 Anton Zeilinger, Markus Arndt and their colleagues in Vienna found a way of altering the rate of decoherence so that they could carry out a detailed comparison of theory and experiment.

They studied interference due to the quantum waviness of entire molecules – a demonstration in itself that quantum mechanics still applies to objects big enough to be seen in a microscope. In the early 1990s researchers devised techniques for making coherent molecular 'matter waves' – quantum-mechanical streams of molecules – which can be used in double-slit interference experiments. The Vienna team reported quantum interference of molecules called fullerenes, containing sixty or seventy carbon atoms joined into closed cages (denoted C_{60} and C_{70}), each almost a millionth of a millimetre across. The researchers marshalled these molecules into a coherent beam that they passed through a vertical grid of slits etched into a slice of ceramic material. Detectors arrayed on the far side of the slits registered an interference pattern: an oscillating rise and fall in the number of molecules detected at different positions.*

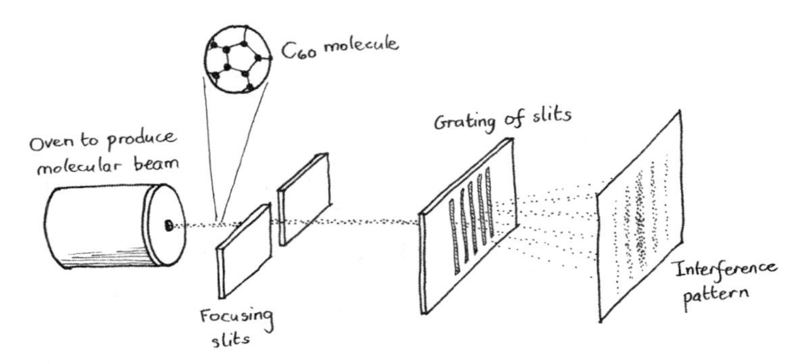

Quantum interference of large molecules (here C_{60}). A collimated (narrowed) beam of molecules is sent through an array of slits, producing an interference pattern due to their coherent wave-like nature. (The detection scheme is considerably more complicated than the simple screen shown here.)

* Interference remains visible for even larger objects. Arndt and his co-workers have subsequently demonstrated it for tailor-made carbon-based molecules of 430 atoms, with sizes of up to 6 millionths of a millimetre – easily big enough to see in the electron microscope, and comparable to the size of small protein molecules in living cells.

The researchers could control the rate of decoherence in these molecular beams by altering the pressure of the gas inside the apparatus: the more gas molecules there are, the more the fullerene molecules collide with them and lose their coherence. As expected, the contrast between the bright and dark interference bands became ever fainter as more methane gas was let into the chamber. This decay of interference reflects the erasure of 'quantumness' in the matter waves due to decoherence. It's possible in this case to predict from quantum-mechanical calculations how strongly a gas of a given pressure should suppress interference. These predictions matched the observations remarkably well, right down to the point where the interference pattern had pretty much vanished. So decoherence not only is real but is accurately described by quantum theory. That theory, in other words, can tell us not just what happens in the quantum world but how quantum becomes classical.

There's more to the story, though. So far all I've outlined here is a theory of what isn't there, namely the coherence we see in quantum mechanics. But to more fully understand measurement and the emergence of the classical world, we also need to understand what *is* there: to explain how decoherence gives rise to the *specifics* that we see around us.

Everything you experience

is a (partial) copy of
what causes it

Decoherence goes some way towards explaining where quantumness goes when a quantum system comes into contact with a classical environment. But a genuine measurement – necessarily classical, involving human-sized apparatus – isn't all about loss. We gain something too: information about the system we're looking at. How is that information related to the properties of the quantum system, and in what ways is it constrained or compromised? How much can we know? Why do classical measuring instruments register the values they do?

So far, in talking about superpositions I have tacitly implied a kind of hierarchy of quantum states. There are states corresponding to the outcomes of measurements, and then there are superpositions of these. The former survive a decohering measurement, the latter don't.

But the whole reason why we can create a superposition in the first place is that it's a valid solution to the Schrödinger equation. Why, then, should *up* and *down* be legal outcomes of measuring a spin, but not *up* + *down*? Why this apparent favouritism? The Schrödinger equation itself doesn't seem to tell us.

You might say, well if we ended up measuring *up* + *down* and so on, our instruments would have needles simultaneously pointing to both positions, and that would be a macroscopic superposition, which is (allegedly) not allowed. But that's no answer: it's just saying that such outcomes are not allowed because we don't see them, and so don't know how to picture them. If the world really was like that, presumably we *would* know how to picture them.

There seemed to be nothing in the theory of quantum mechanics that selects particular quantum states from all the possibilities and says that these and only these correspond to allowed outcomes of a measurement. The theory of decoherence changed that. It can explain why certain solutions to the Schrödinger equation are special – in technical terminology, why there is a 'preferred basis' in quantum mechanics.

And it reveals something truly surprising about how we are able to observe the world.

•

Decoherence is apt to mess things up. If a quantum system is prepared in a superposition, say, decoherence massages and dilutes it until it is unrecognizable in the initial system. But if decoherence did this, in the blink of an eye, to *every* quantum state then we'd never be able to find out anything about a quantum system that hadn't been irreparably corrupted and blurred by the environment. The fact that we *can* make reliable measurements at all is due, first of all, to the robustness of certain quantum states in the face of disruptive decoherence. Some states are special even when immersed in an environment. Wojciech Zurek calls these 'pointer states', because they represent the possible positions of a pointer on the dial of a measuring instrument. Classical behaviour – the existence of well-defined and stable states – is possible only because pointer states exist.

Quantum mechanics lets us figure out what properties they must have. In short, their wavefunctions must possess a certain kind of mathematical symmetry: specifically, decoherence-inducing interactions with the environment simply transform a pointer state into an *identical-looking* state. Recall that the coherence of quantum states is a question of whether the phases of their wavefunctions – the positions of the peaks and troughs, you might say – are aligned. But pointer states are special states for which shifts of phases caused by interaction and entanglement with the environment make no difference. The state still looks the same after the shift. Crudely you can think of it rather like the difference between a circle and a square. You can rotate the circle by any angle you wish and it looks the same, but not so for the square.

This implies that the environment doesn't just squash quantumness indiscriminately: it *selects* particular states and trashes others, a process Zurek calls environment-induced superselection or 'einselection'. The survivors are the pointer states, which are detectable. Superpositions of pointer states do *not* have this stability, and so they are not 'einselected'.

It's not enough, though, for a quantum state to survive decoherence in order for us to be able to measure it. Survival means that the state is measurable in principle – but we still have to *get at* that information to detect the state. So we need to ask how that information becomes available to an experimenter.

Really, who would have thought there'd be so much to the mere act of observation?

•

When we make a classical measurement, we feel that we're directly probing the object we want to investigate. I want to weigh a bag of flour, so I pick up the bag and put it on my scales. It's true that to make this measurement I'm not looking at the bag of flour itself, but at the pointer on the scales. But this doesn't seem a big deal – we know that the weight of the flour is compressing or stretching a spring, and that there's a lever hooked up to that movement which rotates the pointer, or some mechanism of that sort. If you really want to be picky about keeping the measurement *experientially* direct, you can just pick up the bag and, with a bit of experience, make a reasonable estimate of the weight from the downward force on your arm: yes, that feels like about a kilogram.

Oh, but wait: there's a mechanism here too. It just happens to be part of you. There are, in effect, springs and sensors in your arm that register force information and send it to your brain. If your arm was totally anaesthetized, you could hold the bag but no measurement would have been made.

This seems almost idiotically pedantic. Yet we've seen that, in quantum mechanics, this issue of *when the measurement has been made* is crucial to a description of the measurement process. We'd better think the matter through rather cautiously, taking the measurement process step by step.

A measuring device must always have some macroscopic element with which we can interact: a pointer or a display big enough to see, say. As such, it must itself act as a part of the environment interacting with the system we're probing, and so it induces decoherence. That's not wholly destructive, though. Decoherence – entanglement with the environment – is the very process by which *information* passes from the quantum system to its environment. It's what makes this information accessible: what makes the pointer move. Thanks to einselection, the information gets filtered in the process so that only the pointer states survive.

What I'm saying is that *decoherence imprints information about an object onto its environment.* A measurement on that object then amounts to harvesting this information from its environment.

Think of those decoherence-inducing collisions of a dust grain with surrounding air molecules. The changes to the paths of the air molecules due to their impacts on the grain encode a record of the grain's presence and position. If we were able with some amazing instrument to record the trajectories of all the air molecules bouncing off the speck of dust, we could figure out where the speck is *without looking at it directly at all.* We could just monitor the imprint it leaves on its environment.

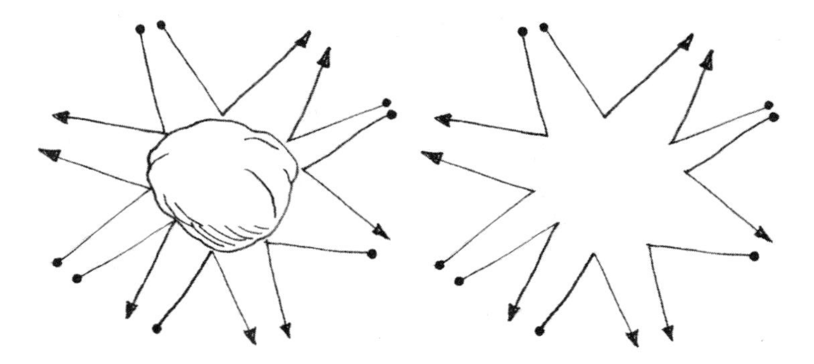

An object's interactions with the environment carry away information about it, such as its position. The location of the dust grain, off which air molecules bounce (left) can be deduced just by looking at those molecules themselves (right). The trajectories of the rebounding molecules encode a kind of replica of the grain.

And this is, in effect, *all we are ever doing* whenever we determine the position, or any other property, of anything. I look down and see my pen on the table – but the only reason I know it's there is because my retinas are responding to the photons of light that have bounced off it. Again this seems trivial – until we recognize that, because of decoherence-inducing entanglement, the information carried away from an object by its environment *fundamentally changes* the quantum nature of the object itself.

This change is *not* necessarily due to any transfer of energy or momentum between the object and the environment, and so it's nothing to do with Heisenberg's notion that observation 'disturbs' what is observed. That disturbance can happen, and indeed it often does: air molecules bouncing off a very small dust grain, for example, transfer momentum to the grain so that it seems to jiggle erratically, because the little impulses it receives from different directions might not balance perfectly. But decoherence doesn't depend on it. Decoherence results from a transfer of quantum *information*: when one object becomes entangled with another, information about each object is no longer confined to the object itself.

The role of decoherence in measurement, then, is not simply to destroy quantum interference and make objects become more classical the more strongly they're wired into their environment. It creates a kind of 'replica' of the object itself – or rather, of the pointer states of that object – in the environment. It is this replica or imprint that eventually produces a reading in our classical measuring apparatus.

We can consider an object's properties to have been measured to the degree that those properties have become entangled with its environment, and have thereby decohered. The stronger the decoherence, the more completely a classical measurement has been made, and the more we have sacrificed the 'quantumness' apparent in the object. It doesn't matter whether the information encoded in the environment is actually read out by an observer; all that matters is that the information does get there, so that it could be read out *in principle*.

So measurement isn't all or nothing; it's a matter of degree. We destroy quantumness in proportion to the amount of information we import from the system into its environment. Zurek and his colleague Bill Wootters have shown that, in a double-slit experiment, it is possible to obtain *some* information about which path a photon took without losing all the quantum interference. While you're not totally certain which path it took, but have reason to think one is more likely than the other, some interference remains. And it turns out that you can get a surprising amount of path information without making the photon completely 'particle-like' and losing all interference. You can become 90% certain about which path while still keeping about half of the contrast of light and dark in the interference bands. (But if you try to get the last 10% of certainty, you'll lose that contrast altogether.)

The information about a quantum object that is carried away by its environment during the process of decoherence is a kind of 'which path' measurement too. The more information the environment 'absorbs' about a dust grain in a superposition of position states, the more the grain becomes localized at just one of those positions, and the less detectable interference there is between them – as we saw for decoherence in the fullerene experiments described earlier.

•

Pointer states don't necessarily get imprinted on the environment in a stable and easily accessible form. Some environments are pretty good at inducing decoherence of a quantum object but not at retaining reliable, sharply defined replicas of it. The collisions of air molecules are like this. Yes, you could reconstruct where an object is from the trajectories of air molecules bouncing off it, but only if you can collect that information before it gets scrambled by the molecules subsequently colliding with one another. Photons, on the other hand, are much better at retaining an imprint, because they don't generally interact with one another after they have bounced off the object and so the information they carry away doesn't get messed up so easily. It's no coincidence that vision is a reliable and widespread way that organisms find out about their environment! Smell, which relies on the passage of odorant molecules through the busy, jostling air, is less good. Some animals use it when vision won't work well (at night, say), but the smeller has to sniff out a wandering, diffusing trail rather than just seeing the target and heading for it.

The efficiency of imprinting replicas also depends on exactly how the system and environment are coupled – which is to say, on exactly how *a measurement is (in principle) made*. In some cases we can calculate, using the equations of quantum mechanics, how efficient this process of replication is. It turns out that some quantum states are better than others at generating 'replicas' – they leave a more robust footprint, which is to say, *more copies*. These are the states that we tend to measure, and are the ones that ultimately produce a unique classical signature from the underlying quantum palette. You could say that it's only the 'fittest' states that survive the measurement process, because they are best at replicating copies in the environment that a measuring device can detect. With that image in mind, Zurek calls the idea quantum Darwinism.

He and his co-worker Jess Riedel have calculated how fast and extensive this proliferation of 'copies' is for a few simple examples, such as a quantum object illuminated by purely thermal radiation (which is what sunlight is, more or less). They find that after being illuminated by the sun for just one microsecond, a grain of dust one micrometre across will have its location imprinted about 100 million times in the scattered photons.

To be observable, then, quantum states need to clear two hurdles. First they need to be robust against decoherence. These are the einselected pointer states, which extract a preferred basis (page 221) and exclude superpositions. Second, they need to be imprinted readily onto the environment. These are the states selected by quantum Darwinism.

Those sound like two separate criteria. But actually they both select the *same* states. That's not a coincidence, though. The robustness against decoherence that defines a pointer state turns out to be precisely what is needed to enable a state to become copied again and again onto the environment without changing it. That property doesn't guarantee that the state *will* be selected by quantum Darwinism – the environment might, for example, simply be very poor at holding onto replicas. But it creates the possibility that, under the right circumstances, the state can be measured.

•

It's because of the multiple imprinting of some states of an object onto its environment that objective, classical-like properties are possible. Ten observers can sequentially measure the position of a dust grain and find that it's in the same location, if no forces act to disturb it between measurements. In this view, we can assign an objective 'position' to the speck not because it 'has' such a position (whatever that means) but because a value of the position can be imprinted in many identical replicas in the environment, so that different observers can reach a consensus about where it is.

In fact, position seems to be one of the most robust properties 'selected' by interactions with the environment. That's simply because those interactions tend to depend on the distance between the object and elements of its environment, such as other atoms or photons: the closer they are, the stronger the interaction. So interactions 'record' position very efficiently. The corollary is that decoherence of position states tends to happen very quickly, because pretty much any scattering of photons from an object carries away positional information into the environment. And so it is really hard to 'see' large-ish objects being in 'two places at once'.

In general, we don't make a measurement by collecting *all* the available information in the environment, but only a part of it. We see objects by looking at some, not all, of the photons they scatter – and that's enough. Quantum Darwinism creates a precise framework for this seemingly (but not genuinely) obvious and mundane fact: it says that the states we can measure are ones that are able not just to imprint themselves in many replicas in the environment, but specifically to do so in many different *parts* of the environment – so that we can find them out without having to look everywhere. The states we can measure are the ones that are most easily found out.

There's a bizarre corollary to this picture. In general, when we measure a property of a quantum system by probing its 'replica' in the environment, we destroy that replica (by entangling it with the measurement apparatus). Might we then potentially 'use up' all the available copies by repeated measurement, so that the state can't any longer be observed? Yes we can: too much measurement will ultimately make the state seem to vanish.

But we needn't be too perplexed by that. What it says is that if we keep poking at a system to find out about it, eventually we'll perturb it into another state. That's completely consistent with our experience. Sure, you can gaze for as long as you like at a coffee mug without altering it in any substantial way. But you can't do that to an Old Master painting, for the pigments will fade under too much light: you will alter their state.

Much less can you examine in prolonged and sustained fashion something small enough that hardly any replicas exist: a single quantum spin, for example. Take a peek and you've used up all the information that was available about it. Subsequent measurements may then have a different outcome. What quantum Darwinism tell us is that, fundamentally, the apparent influence of the observer in quantum mechanics is not really about whether probing physically disturbs what is probed (although that can happen). It is the *gathering of information* that alters the picture. Measurement erases the information that the environment holds about what is measured.

•

There's another profound consequence of this new view of measurement. For a quantum state can never imprint on the environment all that can be known about it in a way that allows this information to be extracted in a single experiment. We can only get a part of the picture with any given measurement. In classical physics this is no big deal, because we can gather all the pieces of the puzzle one by one. We can determine the mass of an object in one experiment, its position in another, its temperature in another and so forth. For a quantum system that piecemeal assembly of the overall picture is no longer possible, because obtaining each piece of information introduces more entanglement that significantly alters some (or all) of the rest – or better to say, it may fix determinate values to properties that were previously indeterminate. Because we can't get all the information out of a quantum system at once, we can't duplicate it exactly. It would be like trying to make a copy of a painting in which the colours get shifted each time we look at it. In quantum mechanics, *cloning is not allowed*. That has some serious consequences, as we'll see.

Here, then, is a concrete way to look at that most puzzling of quantum mysteries, which is that what we measure depends on how we look – on the *context* of the measurement. At face value this jars with our intuitions. But if we have no reason to expect that every possible quantum state leaves its mark on the environment in the same way or to the same degree, and reason to think that the nature of this mark depends on how the system interacts with the environment, then quantum contextuality becomes more understandable, if not indeed inevitable.

Let's be careful about what this means. It's tempting to imagine a quantum object with a whole bunch of properties, some of which produce strong, robust replicas in the environment and some of which imprint only weakly or not at all. And then perhaps a different sort of coupling, a prodding from another direction, is needed to reveal that other information. But this would be to succumb to the habit of realist thinking: to imagine that the quantum object has all its properties intrinsically determined, but that we can only read a few of them at a time. Instead we need to think of the object as having only *potentialities*, which the environment somehow filters and shapes into actualities.

•

If quantum and classical are distinguished only by degree, what exactly is the measure of that degree? John Bell offered one answer: you can look for non-local correlations between entangled states. For quantum states, those correlations can be measurably stronger than anything possible for classical (or hidden-variables) states in which everything knowable about an object is ultimately encoded in that object.

Zurek has offered a more general criterion that extends beyond the EPR-type experiments for which Bell's theorem was devised. The point about a quantum system is that non-local correlations mean you can't know everything about some part of the system by making measurements just on that part. There's always some remaining ignorance. In contrast, once we have established that a glove is left- or right-handed, there's nothing left to be known about its handedness.

By the same token, you *can* potentially find out something about a quantum system by making measurements on another system with which the one of interest is entangled. By looking *here*, you can deduce something about *there*.

That's true too for classical objects whose properties are correlated, like the left- and right-handed gloves. The question is, how much of that information is truly nonlocal: shared between a pair of objects but not deducible by looking at either one of them alone? *That's* where quantumness comes in.

If the two objects are not correlated at all, then you can't deduce anything from the second by looking at the first. If they are perfectly correlated, on the other hand – if, say, they are two gloves identical except for opposite handedness – then you can deduce *everything* about the second by looking just at the first. But in both cases, you gain no more information by looking at the two objects as a pair than you do by looking at each individually.

For quantum systems, though, you *do*. This information that is encoded in the pair but can't be deduced from looking at either or both individually is a measure of their quantumness, and Zurek calls it *quantum discord*. It isn't just a measure of 'how entangled' two quantum objects are: quantum discord quantifies their quantumness even if they're not entangled at all.

You can equivalently think of quantum discord as measuring how much a system is *unavoidably* disrupted – by 'destroying' superpositions or entanglement say – when information about it is gathered by measurement. It's a measure of the ineluctable cost of measurement: how far there is to fall from the misty, elusive quantum heights to the terra firma of the classical valley. For classical systems, the discord is zero. If it is greater than zero, the system has some quantumness to it.

•

We now have, in short, something close to a *theory of measurement*. And it's nothing but the same old quantum mechanics, but now embracing the environment too. It can explain how information gets *out* of the quantum system and *into* the macroscopic apparatus. It allows us (at least in simple cases) to calculate how fast that happens, and how robustly. It explains why some quantities can be meaningfully measured (that is, why they are 'observables' at all) while some can't be. And it is a theory that awards no privileged status to the conscious observer. Measurement now means 'strong interaction with the environment': strong enough, that is, to enable the quantum state to be deduced *in principle* from the imprint it has left, regardless of whether we actually make that deduction or not.

As a result of interaction with the environment, then, quantum coherence is not exactly lost – or rather, it is not lost from the universe. But it becomes invisible to an examination of the quantum system itself, because it is dispersed throughout the environment like an ink droplet spread throughout the ocean. Decoherence means that we can no longer piece together a superposition, much as we can't reconstitute that ink droplet. It's not that the ink no longer exists: the ink molecules are as real as ever. But it isn't very meaningful to think of them as constituting some vast, highly dispersed droplet.

There's then no longer any need for an ambiguous and contentious division of the world into the microscopic, where quantum rules, and macroscopic, which is necessarily classical. We can abandon the search for some hypothetical ''Heisenberg cut' where the two worlds impinge. We can see not only that they are a continuum but also why classical physics is just a special case of quantum physics.

And notice that nowhere in this description have I invoked "collapse of the wavefunction." Does that mean we have done away with that mysterious and problematically non-unitary transformation? Some researchers think so. Given decoherence, says Roland Omnès, to speak of a collapse of the wavefunction becomes 'a convenience, not a necessity'. Sure, he says, we can seek to make wavefunction collapse a real, physical effect by adding extra ingredients to quantum theory. But why bother, if decoherence has already effectively accomplished the same thing?

But most quantum researchers don't agree. The problem that obstinately remains is, in a nutshell, uniqueness. Quantum mechanics offers us many possibilities – many potential realities. As they become entangled with their environment, the options are boiled down: classical states emerge purely and simply by quantum mechanics doing its thing. That's something of a revelation; it relieves us of the need to treat the large and the small as alien to one another.

But there's still one more step involved in making a measurement. Superpositions of quantum states are replaced by well-defined classical states – but *we only see one of them*!

How does that selection happen? Why, in any given measurement, do we see *this* and not *that*, when both *this* or *that* (though no longer '*this and that*') are classically possible? Where did the other possibilities go? Sure, one might say that they were dispersed into the surroundings – but like the ink molecules in the dispersed drop, they are then still in principle *out there*. Why can't we find them? Perhaps we *can* still find them?

Well, maybe – but not according to conventional quantum mechanics. For the way the theory is used to deduce the outcome of a measurement still demands, in the final step, the mathematical transformation that some label 'wavefunction collapse', by which multiple possibilities become a single actuality. If you want to connect to a world that yields only unique experiences, you still need collapse. And then you must ask: so what is this thing we call 'collapse'? Is it just an updating of our knowledge about the system (as an epistemic view might imply), or of our *beliefs* about likely measurement outcomes (as a QBist would say), or an actual physical process, or an axiomatic aspect of the theory that you'd better accept with no questions asked, or . . .?

Decoherence theory can tell us an awful lot about how quantum becomes classical–about how the counter-intuitive aspects of quantum rules become classical "common sense". But it can't finally get us to that most common-sensical feature of all: why *this*, not *that*? *Why are there facts of the world?*

Schrödinger's cat

has had kittens

We need to talk about that cat.

Did decoherence kill it – or conversely, sustain it? The environment, it seems, will 'measure' the cat whether we like it or not. If it's to be alive at all then it must be surrounded by colliding air molecules and bathed in thermal photons, which will suffice to place the cat in a classical state (dead or alive) whether we open the box or whether we do not.

But that doesn't quite answer the question. There's nothing to prevent us from suppressing decoherence in principle, even if we can't realistically do it in practice. Give the cat an oxygen mask and a thermal suit and suspend it in an ultracold vacuum, or whatever ridiculous extremes your thought experiment demands. What then?

At face value, quantum mechanics insists that a superposition of live and dead states should then be possible. Some researchers today are happy to accept that: to assume that there need be nothing so absurd about live/dead cats. They don't feel obliged, as Schrödinger and Einstein did, to regard such a surreal prospect as inherently absurd.

In truth, the question has little real meaning unless we can define 'live' and 'dead' in quantum terms, so that we might actually write the wavefunction of a superposition of the two states and calculate how it evolves. And it's simply not clear how to do that – it's not a sufficiently well-defined scenario. Arguably we should let the matter drop there.

Yet that won't quite get us off the hook, because as we saw, the cat is superfluous anyway. Schrödinger wanted to highlight macroscopic states that are mutually exclusive by definition, so as to draw out the logical contradictions that appear to loom in large-scale quantumness. But we might equally think about superpositions of large objects more amenable to the theories of physics. A coexistence of two different positions, say, is hard to imagine intuitively, but two such states are not in semantic opposition and are much more readily specified and measured: you just need to find the object's centre of gravity. A (live) cat *here* and the same (live) cat *there* is a strange thing to think about, but personally I don't struggle with it in quite the same way as I do with Schrödinger's juggling of life and death.

Creating that situation in reality with something large, warm, furry and mobile is another matter. But something a little smaller and well behaved, perhaps? Might we then control its interactions with the environment to suppress decoherence? Experimental scientists are now striving to produce superpositions, interference and other quantum phenomena in middle-sized, mesoscale objects, to put to the test what quantum mechanics now seems to tell us: that the quantum–classical boundary is merely a practical limitation and not a fundamental one. That it's just a problem of engineering.

Rightly or wrongly, they call such mesoscale systems 'Schrödinger's kittens'.

•

No living thing is smaller than viruses. Particles made simply of DNA or RNA packaged in a protein coat and primed to replicate inside a host organism, they can be as small as twenty nanometres (millionths of a millimetre) across. There is still debate about whether they qualify as genuinely 'living' entities, but they are undoubtedly part of the biological world. Might we, then, make Schrödinger's viruses?

That idea has been proposed by Ignacio Cirac and Oriol Romero-Isart at the Max Planck Institute for Quantum Optics in Garching, Germany. They have outlined an experiment for preparing not only viruses but also extremely hardy microscopic creatures called tardigrades or water bears – which are up to a full millimetre or so in size – in superposition states. Tardigrades can survive on the outside of spacecraft beyond Earth's atmosphere, and so they might withstand the high vacuum and low temperatures needed to suppress decoherence.

The idea is to levitate the organisms in an 'optical trap', which uses intense laser-light fields that create a force keeping the object in the brightest part of the beam. The objects would vibrate in the trap as if they were suspended from springs. The aim is then to manipulate the trapping forces so as to coax the objects into a superposition of vibrating states: vibrating, say, 1,000 and 2,000 times a second. A simple way to look for quantum behaviour would be to make the states interfere and search for signatures of that interference.

Achieving this kind of superposition with living objects has no fundamental significance in itself. You might just as well do it with grains of granite. It is not as if a tardigrade can tell you what it feels like to be in a superposition. All the same, a demonstration of this sort would offer compelling evidence that (as most scientists already suspect) life itself is no obstacle to detectable manifestations of quantum mechanics.

•

On the whole, Schrödinger's kittens tend to be soberly inanimate. Some researchers hope to coax them from tiny, springy structures called nanomechanical resonators: microscopic cantilevers, beams, and thin, stiff drumskin-like membranes. A typical nanomechanical resonator is a beam of material several micrometres long and a micrometre or so wide, fixed at both ends across an empty space like a tiny bridge. These structures have particular resonant frequencies, but because they are so small each resonance is governed by quantum rules: the amount of energy it may contain is restricted to particular quantized values. The smaller the structure, the more widely separated and readily distinguished their quantum energy states are.

A microscopic resonating 'springboard' used in attempts to create quantum vibrational states at the 'mesoscale'. The springboard is about as long as the width of a human hair, and is twisted because of stresses in the materials. Image courtesy of Aaron D. O'Connell and Andrew N. Cleland, University of California, Santa Barbara.

To place such an oscillator in a superposition of vibrational states, you first need to get it securely under control: to prepare it in the lowest energy state, called the ground state. Heat is apt to excite higher-energy vibrations too, so these nanomechanical resonators must be made very cold indeed. They can be chilled close to absolute zero using cryogenics, and then cooled some more by using laser beams to soothe the remaining vibrations – a technique called laser cooling. In this way, the tiny vibrating objects can be guided into a single quantum state.

The experimenter then has to place the tamed resonator in a superposition. One way to do that is to hook it up to another quantum object whose state can be easily manipulated. The ideal control system is a 'quantum bit' or qubit, an object that can be switched between two clearly distinguished quantum states, such as an atomic spin orientated *up* or *down*. A qubit doesn't have to be in one of these states or the other, but can exist in a superposition of both.* If the resonator's state is controlled by the qubit, the resonator too can then be placed in a superposition.

These experiments need extraordinary sensitivity, because they are looking for very small effects in comparatively big things – rather like trying to detect the vibrations of the Golden Gate Bridge due to a bicycle pedalled over it. Andrew Cleland at the University of California at Santa Barbara and his co-workers have managed to couple a resonator made from a microscopic sheet of a hard ceramic

* Remember that a superposition is not really 'two states at once', but a circumstance in which either state is a possible measurement outcome.

material to a qubit made from a ring of superconducting material (page 264). They hope to prepare two resonators in an entangled state and then to probe the correlations between the vibrations of the sheets, in an experiment a little like those used to look for violations of Bell's condition for EPR correlations. Other researchers are trying to prepare individual oscillators in superposition states and watch how they decohere as they get entangled with their environment: middle-sized Schrödinger kittens leaking quantumness into the surrounding space.

•

Most researchers anticipate that these studies of Schrödinger's kittens will reveal that the sole limits on observing quantum-mechanical behaviour stem from the difficulty of suppressing environment-induced decoherence. In that view, the only significance of the kittens' physical size is to make decoherence harder to avoid.

But it's possible that the emergence of classical behaviour involves something more too. Physicists Johannes Kofler at the Max Planck Institute for Quantum Optics and Časlav Brukner of the University of Vienna think that even if decoherence can be suppressed, we might only ever be able to see classical behaviour in a large object. They think that inevitability could be a consequence of the limited precision of measurements, in which there's always a margin of uncertainty or inaccuracy.

The argument often made in textbooks is that such limits on experimental resolution prevent us from being able to see quantum discreteness in a macroscopic system because the discrete energy states get ever closer as the size of the system increases. The states then seem to blur into the continuum of energies that we perceive in, say, a moving tennis ball. But that can't be the whole story, because it doesn't actually eliminate the quantumness of the object – forbidding, for example, a superposition of tennis-ball velocities. It just means quantumness gets extremely fine-grained.

Kofler and Brukner argue, though, that the 'coarse-grained' nature of measurement – meaning that the instrumental resolution is unable to distinguish several closely spaced quantum states of a large system – makes this quantumness mimic classical physics. Filtered through such a coarse-grained lens, the quantum-mechanical equations describing how such a large object changes through time are reduced to Newton's classical equations of mechanics, washing out non-local features such as entanglement.

In other words, classical physics emerges from quantum physics when measurement becomes imprecise, as it always must for sufficiently big systems. It's not that the quantum coherence goes away; rather, it just can't be seen any more. And it's not just that we can't then tell the quantum states apart; the *specific physical laws* that emerge from quantum mechanics in those circumstances are precisely those of classical physics. Once again, the classical world is simply what quantum looks like when you're human-sized.

This picture offers an alternative resolution to the puzzle of Schrödinger's cat. We could never see it in a live/dead superposition not because that doesn't exist or because decoherence collapses it (although that can happen too) but because, well, we just couldn't make it out: our instruments lack the necessary precision.

But surely you don't need some incredibly delicate instrument to tell a live cat from a dead one! Remember, though, that this isn't the aim. You're not asking 'Is it alive or dead?' If the cat gets up and licks at a bowl of cream, the superposition has already collapsed. Indeed, you won't see a superposition at all just by looking: all those photons bouncing off it are sure to induce decoherence. No, remember from the double-slit experiments that the way we detect a superposition is precisely by *not* making a measurement of the state (did the photon pass through *this* slit or not?), but instead by looking for interference between the superposed states.

And how do you do that for a live/dead cat? What are you going to measure? Heartbeat? Find one, and the cat's alive. Temperature? That won't even distinguish a live cat from a just-dead cat, so it's not clear how it can reveal a superposition. As I said earlier, 'live' and 'dead' are not well-defined quantum (or perhaps even classical) states, so we don't know what to measure.

Even if we decide instead to place the cat in a positional superposition rather than a live/dead one – because maybe we *can* measure interference between positions (though I'm not sure how you'd achieve that for a cat) – then we can't hope to detect it. The only kind of positional superposition we could imaginably sustain even for the briefest instant for a big, warm object like this is one in which the two positions are too close together for interference between them to show up within the resolution limits of any plausible instrument. What we end up seeing obeys Newton's laws, not Schrödinger's.

This emergence of classical physics through the blurred lens of measurement is an idea amenable to experimental testing. It would mean creating a largish Schrödinger-cat-like state that still retains demonstrable quantum behaviour (such as interference), and then looking to see if this goes away as the precision of the measurements gets steadily coarser. In principle it could be done, but it won't be easy.

•

What is it, in the end, that distinguishes a quantum from a classical object? Well, isn't that obvious by now? Classical objects can't be in superpositions (crudely, 'in two states at once'), they can't be entangled, they don't show wave-like interference. But this amounts to saying that classical objects don't exhibit certain types of experimental behaviour. It tells us what to look for; but what is the *fundamental* difference?

Our classical preconception that objects have properties firmly located on themselves is called *local realism*. Not only are these properties *local* (they're not affected by things that are too far away to interact while we're measuring), but they are *real* in the sense of being pre-existing and amenable to confirmation. Different observers can probe the same object and agree on what it is like – not just because they happen to measure the same values by chance but because those are the values intrinsically associated with the object in question.

In 1985, physicists Anthony Leggett and Anupam Garg proposed some ground rules for what they called *macro-realism*: the idea that objects will behave in this 'realistic' way we have come to expect in the macroscopic world. Leggett and Garg worked out what kind of observations would be compatible with such rules. They expressed this compatibility as a limit rather like that which Bell's test imposes on the measurement of correlations between particles described by hidden variables. If experiments violate the Leggett–Garg limit, the objects being probed are not macrorealistic.

Over the past four years, several experiments have shown that Leggett–Garg macrorealism is indeed violated for relatively small systems – as we would expect when quantum rules apply. The question is whether such violations remain possible as the size of the systems increase. The problem is that the experiments then get progressively harder. And so we don't yet know if, say, a peanut can violate macrorealism, if we were clever enough to create conditions that make it possible.

The Leggett–Garg criterion isn't predicated on counter-intuitive quantum effects, but is approaching the issue from the other direction: asking how far we can rely on our regular experience of the world. *All* physical theories that respect locality and realism (like Newtonian mechanics) would stay within the Leggett–Garg bounds. So it isn't so much a test of 'how far up' quantum behaviour can reach, but of 'how far down' the characteristic features of our classical world extend – if indeed they are a fundamental requirement at all.

•

Suppose that what we take to be macrorealism turns out to be an illusion orchestrated by decoherence, and that *in principle* quantum effects such as superposition really can exist at all scales. Might we then ever find ways to grow Schrödinger's kittens all the way up to cats, and to *see directly* how they sustain the potential to exhibit more than just one distinct value of some property in a measurement? The technical challenges of suppressing decoherence are enormous, and might never be overcome. But all the same it doesn't seem obviously futile to wonder what a macroscopic quantum phenomenon would *look like*.

In one sense, we can see them already. Superconductivity – an ability to conduct electricity without any electrical resistance – is a property that quantum effects confer on some materials (such as metals) at very low temperatures. And when a material superconducts, a magnet can be levitated above it, kept visibly aloft by quantum mechanics in action. Superfluidity, another quantum effect, allows ultracold liquid helium to flow, like some science-fictional gloop, up the sides of a container and out of the top. You can see those strange things happening with the naked eye. Yet, however bizarre and impressive they look, they are not 'quantum' in the sense we've been talking about so far. They are large-scale consequences of recondite underlying quantum principles. They are not 'two states at once'.

Tiny mechanical beams vibrating in a superposition of states are unlikely to satisfy a desire to see the spine-tinglingly odd either. In general we would only detect these things indirectly, not see them with our unaided eyes. Even if we were able to image them with a microscope without disrupting their delicate quantum states, it seems unlikely that we'd notice anything untoward – the effects are too slight.

But some researchers hold out the hope that superpositions of photon states might impinge directly on our consciousness, thanks to the extraordinary sensitivity of our visual system. The light-sensitive rod cells in our retinas that register low levels of illumination – they are responsible for our night vision, taking over from the regular cone cells as twilight falls – are incredibly sensitive photon detectors. Researchers at the University of Illinois at Urbana-Champaign have shown that these cells can detect pulses containing as few as three photons. They placed volunteers in a darkened room and exposed them to flashes of light each composed of just thirty or so photons, produced by modern optical devices that deliver individual photons to order. The participants *think* they see nothing, but they're told that the flashes are there and are asked to make guesses about whether they come from the left or the right. They guess correctly more often than can be ascribed to sheer chance. Because the eye as a whole is not a perfectly efficient photon detector, at least 90% of the photons in these flashes will be absorbed before they reach the retina. This means that on average, only three photons hit the rod cells each time.

What might happen, then, if the photons in the flashes are placed in a superposition of states? How will that affect what the experimental subjects 'see'? Would it set up some kind of superposition in the nerve impulse from the rod cell to the brain? Might it even create a superposition of *perceptions*? It seems rather likely that, if the experiment is ever performed (that hasn't happened yet), the result will be not some strange, novel state of mind but just more of the same, since the rod cell will act like any other macroscopic measurement device to transform a quantum state to a classical one, decohering it in (literally) a flash. But right now, no one knows.

Quantum mechanics can be

harnessed for technology

Making collections of many quantum particles suspended in entangled or superposition states – embryonic Schrödinger kittens, if you will – is more than a matter of academic curiosity. If we master the art, we can use these quantum collectives to do useful things. We can, for example, make computers that run on quantum rules.

Quantum computers already exist, and the first to be made commercially available certainly looked the part. Called D-Wave and marketed by a company of that name in Burnaby, British Columbia, it's a mysterious black box straight out of science-fictional central casting, the size of an industrial refrigerator (but considerably colder inside). At $10 million a shot, D-Wave isn't exactly a consumer item, but technology giants such as Google, NASA and the aerospace and advanced-technology company Lockheed Martin have all bought one.

Truth be told, whether D-Wave really is the world's first commercial quantum computer is disputed, for it works in a different way to most quantum computers under development. But prototype quantum computers of a somewhat more mainstream design have now also been produced by IBM and Google, and there's a fair chance that others will appear between the completion and publication of this book.

Quantum computers harness the principles of quantum mechanics to greatly accelerate the rate at which they process information. Ultimately they might do in seconds computations that classical machines would labour over for weeks or even years, because quantum computers can process information in ways that are fundamentally impossible on classical machines.

It's not clear, though, if quantum computers will ever supplant your laptop, even if the current price tag falls. In theory quantum computation should be phenomenally good at solving certain kinds of problems, but we don't yet know if there's much to be gained from using their quantum tricks for *all* computing. One of the big challenges in the field is not just building these machines but figuring out ways of putting them to good use.

All the same, the mere existence of even the most rudimentary of quantum computers demonstrates that quantum mechanics has moved far beyond being a language to describe an esoteric world that most people never encounter. The use of quantum mechanics to improve information technology supplies one of the most compelling demonstrations that it really does describe *something* real about the world.

The significance goes deeper, however. The very idea of treating quantum systems as repositories of information, which can be stored, manipulated and read out just as it can using the digital circuitry of conventional computers, reinforces the view that information lies at the heart of the theory. It's for this reason that quantum computing is more than a practical offshoot of the physics of quantum mechanics. It speaks to some of the foundational issues of the discipline. What is possible and what is impossible in quantum computing follow from the same rules that govern what is knowable and what is not.

Quantum computing is, then, a two-way street. It exemplifies not so much the common (but misleading) notion that basic science leads to applied, but the fact that the rigorous demands of technological application force 'pure' science to confront what it does not know, and perhaps to advance and refine that knowledge. Many of the pioneers of quantum computing are the same folk who think most profoundly about what quantum mechanics means. Had these machines and related quantum information technologies been invented sooner – and really there is no clear reason why they should not have been – we can be sure that the likes of Bohr, Einstein, John von Neumann and John Wheeler would have had plenty to say about them.

•

After all, one of those quantum pioneers is credited with the initial concept. In 1982 Richard Feynman wondered about the best way of 'simulating physics with computers'. Computer simulation is now a mature discipline: a way of predicting how things behave by representing them as a kind of computer model governed by physical laws, and letting the laws unfold to see what emerges. The equations are typically simple in themselves, but there are an awful lot of them, and we must solve them over and over again at each instant of the process we're modelling. So we let the computer do it, because computers are much faster and better at that kind of thing.

Computer simulation often works fine if we assume nothing more than Newton's laws at the atomic scale, even though we know that really we should be using quantum, not classical, mechanics at that level. But sometimes approximating the behaviour of atoms as though they were classical billiard-ball particles isn't sufficient. We really do need to take quantum behaviour into account to accurately model chemical reactions involved in industrial catalysis or drug action, say. We can do that by solving the Schrödinger equation for the particles, but only approximately: we need to make lots of simplifications if the math is to be tractable.

But what if we had a computer that itself works by the laws of quantum mechanics? Then the sort of behaviour you're trying to simulate is built into the very way the machine operates: it is hardwired into the fabric. This was the point Feynman made in his article. But no such machines existed. At any rate they would, as he pointed out with wry understatement, be 'machines of a different kind' from any computer built so far. Feynman didn't work out the full theory of what such a machine would look like or how it would work – but he insisted that 'if you want to make a simulation of nature, you'd better make it quantum-mechanical'.

Feynman wasn't thinking about doing computing *faster*. Rather, he imagined that a quantum computer could do things that were simply impossible for classical machines. Some researchers continue to think that it will be this aspect, and not 'quantum speed-up', that will furnish the best justification for the immense effort being devoted to making quantum computers. Perhaps the focus on speed, particularly in media reports, reflects our experience of personal computing. When Feynman was writing, few people imagined how pervasive computers would become in daily life, or how critically dependent on speed that role would be. Today a claim that a computer will be faster than what came before needs no further justification of its value.

In any case, despite all the noise about how they'll calculate faster, until very recently no one had built a quantum computer that can do much more than the kind of sums a school child (let alone a classical computer) can perform with ease. The obstacles to tackling tougher questions could be regarded as just a matter of engineering. But in fact both the challenges and their potential solutions depend on fundamental features of quantum physics, and it's unlikely that they will be solved without a better grasp of those principles.

•

All of today's computers use binary logic, encoding information in strings of 1s and 0s. These binary digits (bits) may be represented by, for example, electrical pulses in wires, or by flashes of light in optical fibres, or by the orientation of magnetic poles in some memory device. The physical implementation doesn't matter, and one form of encoding can be converted to another (as it is when data is stored or transmitted) without changing the information itself.

The logic operations of computation interconvert 1s and 0s according to certain rules. A 'logic gate' receives input signals – a 1 and a 0, say – and combines them into output signals. An AND gate, for instance, gives an output of 1 only if both of its two inputs are 1, and outputs a 0 for any other combination. In conventional microprocessor circuits these gates are built mostly from transistors made of silicon (and related semiconducting or insulating materials), which act as tiny switches. Carrying out a particular computation involves enacting a certain sequence of steps – an algorithm – that combines and manipulates input data so as to transform it into the solution to the problem at hand. Different computations deploy different algorithms.

This manipulation of bits is really all that computation – *any* computation – is. The rest is a question of building software and interfaces that turn these bits into symbols glowing on the screen, or ink sprayed onto paper, or whatever it takes for us to be able to communicate with the machine and vice versa.

Quantum computers use 1s and 0s too, but with a crucial difference. Their fundamental trick is to replace classical bits with quantum bits (qubits), in which the binary information is encoded in quantum states. These states could, for example, be the two polarization states of a photon, or the *up* and *down* spin states of an electron or an atom.

As we saw earlier, qubits can be placed in superpositions of states, encoding not just a binary 1 or 0 but any combination of the two. A qubit can be considered to simultaneously encode a 1 and a 0, or 1 with a tiny bit of 0, and so on. While a classical bit exists in only two different states, qubits can access a vast range of states, that vastness expanding rapidly as the number of qubits increases. Because of this widening of options, you can manipulate information much more efficiently in an array of qubits than in an array of classical bits.

Somehow – I will look later at just how this comes about, the spoiler being that we don't fully know how – this facility with information can make a quantum computer much faster at solving some calculations than a classical computer could be. The aim is to perform logic operations using qubits that interact so that the information they encode is shuffled into new configurations while maintaining its quantum-mechanical character – which means keeping the qubit superpositions *coherent* (page 204). Just as in a classical computer, the 1s and 0s of the input to a quantum algorithm are marshalled into binary digits encoding solutions.

The catch is that superpositions are generally very 'delicate'. They get easily disrupted by disturbances from the surrounding environment, particularly the randomizing effects of heat. As we saw earlier, this doesn't really mean – as is often implied – that superpositions are destroyed, but rather that the quantum coherence spreads into the environment, so that the original system decoheres. Once decoherence happens, the qubits are scrambled and the computation collapses. Crudely speaking, their 1s and 0s are then no longer all part of the same message.

This needn't be an all-or-nothing affair. Rather, it may be that environmental disturbances such as heat flip just a single qubit from its intended quantum state into the alternative state, making a 1 into a 0, say. Then the computation can proceed, but it has been corrupted and the answer might be unreliable.

In general, a bunch of qubits is collectively stable only at very low temperatures, where the errors introduced by thermal noise are kept minimal. This fragility of quantum coherence in an array of qubits means that, although the theory of quantum computation is already well advanced, building a practical device is taxing the skills of electrical and optical engineers and applied physicists to the limit. It's not yet feasible to assemble more than a handful of qubits and keep them in a superposition long enough to do any computing with them. So quantum computers have not yet achieved anything that couldn't be done with considerably less effort on a classical computer.

•

Following Feynman's prescient suggestion, the theory of quantum computation was developed in the mid-1980s by David Deutsch at the University of Oxford, Charles Bennett at IBM's Yorktown Heights research laboratories in New York, and others. But it took several years before anyone figured out an algorithm for manipulating qubits that could achieve something useful.

In 1994, mathematician Peter Shor at the Massachusetts Institute of Technology devised a quantum algorithm for factorizing large numbers: breaking them down into divisors that are prime numbers, which by definition can't be divided any further. For example, 12 can be factorized into $2 \times 2 \times 3$, while the prime factors of 21 are 7 and 3. There's no known shortcut to finding the prime factors of a number: you just have to try all the possibilities. For example, that the prime factors of 1,007 are 19 and 53, while 1,033 has none (it is itself a prime number), are things you'd only find out by trial and error.

So factorization demands a laborious search through all the possible answers, one by one. That's how a classical computer would do it. Because it can perform simple arithmetical calculations at lightning speed, it can generally find factors for numbers like this in an instant. But as the numbers get bigger, the number of calculations a computer must plod through spirals quickly. To factor a 232-digit number, hundreds of computers took two years before finally delivering a result in 2009. Finding the factors of a 1,000-digit number with today's machines is simply not practical: it would take many lifetimes.

This difficulty of finding factors of large numbers is used for data encryption. If the data is encoded in a way that can be cracked only by solving a large-number factorization problem, it can't be decoded on any reasonable timescale even by a supercomputer. If you're given the factors, though – the key to the code – then decoding becomes very easy.

The length of time it takes to factorize some number *N* classically by searching through the possibilities increases exponentially – which is really just to say, it gets rapidly longer – as *N* gets bigger. But Shor showed that with a quantum algorithm you could find the factors in a time that increases with the size of *N* substantially less fast than that. In other words, although the problem still takes ever longer as *N* gets bigger, it doesn't take *as* long as it does on a classical computer. If Shor's algorithm could be implemented on a big enough quantum computer, it would be able to crack all the current data encryption codes based on factorization.

That's not going to happen soon. Because of the technical challenges of making a real quantum computer, Shor's algorithm has only so far been implemented in a few qubits – sufficient to factorize the number 21, say. Even doing this much, which would (one hopes) take a school child a few seconds, involved a tour de force of quantum engineering. Other quantum algorithms for factorization have now been devised, and one has factored considerably larger numbers than has Shor's – but still they have attained nothing that would make your laptop break into a sweat.

Another task that lacks any efficient means of solution beyond trial and error is the searching of large databases – say, to match a stored record with the one you have in your hand. You have little choice but to look at each item in the database in turn – as it were, to search through every drawer. This means that the time taken to find what you want gets larger in direct proportion to the number of items you have to sort through. In 1996,

Lov Grover of Bell Labs in New Jersey reported a quantum algorithm that could search much faster: the time taken gets longer only in proportion to the square root of the number of items. So for a database with a hundred items in it, Grover's algorithm can conduct the search in a tenth of the time it takes a classical computer. The slowness of this kind of classical search is also used as a basis for data encryption, and so Grover's quantum algorithm again possesses code-cracking potential.

The quantum algorithms of Shor and Grover demonstrated that quantum computers might carry out calculations faster than classical ones, shifting the emphasis of the field from what new things one could do with a quantum computer (like simulating nature very accurately at the atomic scale) to how fast it could work. Yet for all the much-vaunted benefits of quantum speed-up, it remains a challenge to find problems, besides factorization and searches, that quantum computers would solve faster than classical machines. There are so far rather few tried-and-tested quantum algorithms, and some researchers think that quantum computers might prove to be niche devices: brilliant at some things, no better than their classical counterparts at others.

•

Questions about what quantum computers might do haven't held back efforts to make them. The first question is how to make the qubits, and how to couple them together so that they may sustain coherent superpositions. Remember that a superposition of two or more particles corresponds to an entangled state, so quantum computers in general require the qubits to be entangled.

Entanglement is not an *essential* requirement, as we'll see, but most proposals for quantum computing rely on it.

To keep entanglement localized in the qubits – to prevent it from decohering – these elements must be isolated from their environment as much as possible, while retaining the capacity to feed information into them and read it out. One idea is to encode data into the quantum energy states of atoms or ions (electrically charged atoms) trapped using light or electromagnetic forces. Alternatively we could use atomic spins as the qubits, for example by implanting atoms with spin in some kind of matrix: impurities embedded in a solid material such as silicon, like raisins in a cake. The most promising qubit is proving to be a ring of superconducting material, in which bits may be encoded in the direction in which an electrical current circulates. Typically, superconducting qubits can hold on securely to their data in the face of thermal noise only if they are cooled to within just a few thousandths of a degree of absolute zero.

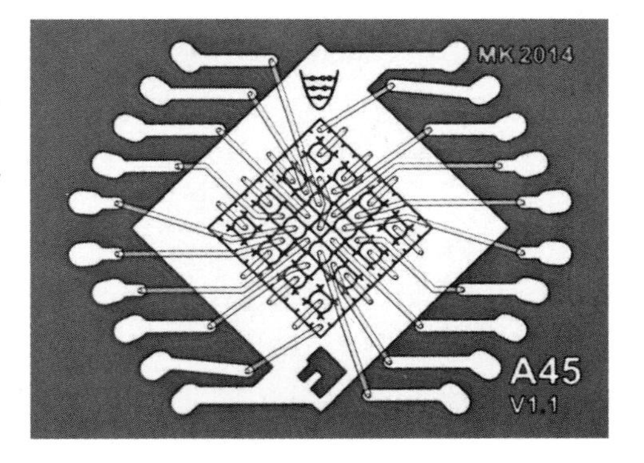

An array of electrical traps on a chip about a millimetre across, built to hold ions for quantum computing at the University of Innsbruck in Austria. Courtesy of M. Kumph, Ph. Holz, K. Lahkmanskiy and S. Partel/University of Innsbruck and FH Vorarlberg.

The quantum computers made by D-Wave, IBM and Google all use superconducting qubits. The IBM device is the most conventional: a microprocessor comprising five digital qubits. In 2016 the company launched a cloud-based platform that allows public users to sample the capabilities online. At the time of writing, IBM and Google are unveiling devices with 49–50 qubits. These figures don't sound very impressive in comparison with the billions of bits in today's laptops. What's more, effective computation requires many qubits to be assembled into each single *logical* qubit, which has the full capacity – in particular the ability to correct random errors – needed for logic processing. All the same, it will take only forty to fifty logical qubits for a quantum computer to outperform the best of current classical supercomputers for certain tasks – an achievement grandiosely (if not indeed rather ominously) called 'quantum supremacy'.

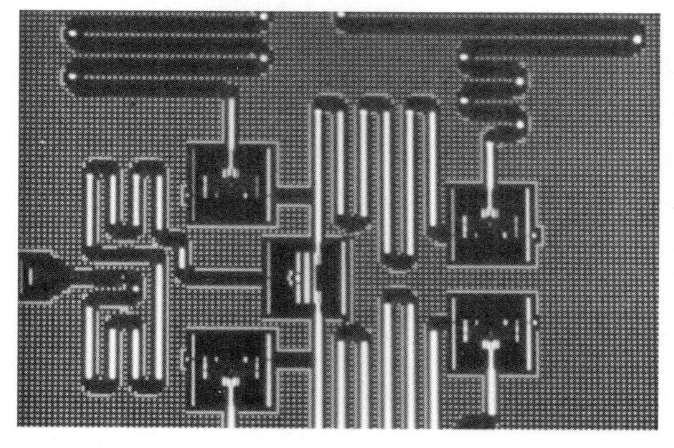

IBM's 5-qubit quantum microprocessor, unveiled as the cloud-based IBM Quantum Experience in 2016. Each of the square structures is a superconducting qubit. Courtesy of IBM.

The most taxing computational problems are currently tackled by massive, phenomenally expensive (classical) supercomputers housed in a few specialized centres and leased to users. The initial market for quantum computers will look like that too: not a real market at all, but a highly centralized oligopoly. But of course that's how *all* computers used to be: huge mainframes used by an elite and dedicated to recondite problems. Mindful of IBM founder Thomas Watson's (apocryphal) prediction in 1943 that the computational needs of the world would be satisfied by just five of these monsters, it would be a bold or reckless prophet who forecasts where quantum computers might sit a few decades from now.

•

One of the biggest problems for a quantum computer is how to deal with errors. Even a quantum computer will get things wrong occasionally: a 1 adventitiously switching to a 0, say.

That happens in classical computers, but it's not hard to cope with. You just keep several copies of each bit, updating them all whenever necessary. If you have three copies and one differs from the other two, you can be pretty sure that it's been changed by mistake – by some randomness in the circuitry – and can correct it.

This checking and correcting is essential, because otherwise errors can accumulate and propagate, just as they do in a school math problem: one slip in your workings and it all goes haywire from that point on. But in quantum computing, this method of error correction by keeping redundant copies of the information isn't possible. The problem is that, in general, if you do something to a quantum state, you can only get a *different* state. You can't get a duplicate of the original state.

This is a fundamental aspect of quantum mechanics that we encountered previously, called the no-cloning principle: *you can't make an exact copy of any arbitrary (unknown) quantum state.*

The 'quantum no-cloning rule' is something of a misnomer, because copying a quantum state is not absolutely forbidden. You can do it for certain kinds of state under special circumstances. Never mind what these are; the key point is that, to do so, you need a copying device tailor-made for the job. It will only work for the state for which it was designed. So you can never clone an arbitrary, *unknown* quantum state, because you wouldn't know which kind of copier to use.

Quantum no-cloning might seem like a mere technical inconvenience, but it's actually a deep principle. For one thing, if exact copying were possible, then it could provide a means of sending information instantaneously over long distances using entanglement. So you could say that the prohibition on cloning is a way of safeguarding special relativity.

But at the root of no-cloning is the fact that an 'unknown quantum state' is not like, say, an unknown telephone number. It's not merely something we don't know about; it's something that, because it has not been observed in some way, *has not yet been determined*. Indeed, if we take the epistemic view that a quantum state reflects a state of knowledge about a system, then an 'unknown quantum state' is an oxymoron: if there's no knowledge, there's no state. So there's a close connection between trying to copy a quantum state and trying to measure it. You can't do either without leaving the original state altered in some way.

Think of it like this. Both measurement and (attempted) duplication of a quantum state must follow rules in which only certain outcomes are permitted. You can't ask a quantum state 'What state are you?', but only 'Are you *this*? Are you *that*?' And you then get only the answers 'Yes' or 'No' – but in doing so you risk scrambling the answers you *would* have received to other questions. Only if you know the state in advance can you know exactly which are the 'right' questions to ask, so that you can avoid messing up some of the answers. In other words, quantum no-cloning is a consequence of the fact that you can't find out in a single shot everything *potentially knowable* about an arbitrary quantum state.

Where does this fundamental limitation on probing and manipulating quantum information leave the quantum computer? At first blush, it doesn't look too promising. Not only can we not generally copy a qubit to provide security against errors, but neither can we look to see if an error has occurred in the first place, without in both cases actually *measuring* the qubit and thus destroying the all-important superposition or entanglement on which quantum computation depends.

When quantum computing was still just an idea on paper, it seemed that the problem of error correction might be a fatal flaw. But from the mid-1990s researchers figured out methods for detecting, correcting and suppressing qubit errors. The trick is to figure out if the value of the qubit has changed from what it should be, without actually 'looking' at that value. One strategy uses redundant data encoding, somewhat like that in classical computing but with greater cunning. Extra qubits are included in the system which are not actually necessary for the computation but which are coupled to the ones that are, so that they have inter-dependent values. With a bit of clever planning, these so-called ancilla ('servant' or 'slave') qubits can be interrogated to discover if any errors have appeared, without finding out any specific information about the states of the principal qubits themselves. By getting the information at one remove, you might say, it is possible to 'deny' ever having seen it.

The ancilla qubits may then act as a handle that can be manipulated to nudge the main qubits back into the right state (or close enough) without doing anything to them directly. Again, by relaying the instruction indirectly, one can 'deny' ever having given it.

Researchers have also sought methods of coding and processing quantum information that suppress errors in the first place. Alternatively we can deal with error by learning to live with it: to find ways of carrying out a quantum computation that is very tolerant of flaws. It's clear enough in principle that a few errors needn't totally wreck a computation. If, for example, a few votes in a general election are miscounted (indeed, they inevitably will be), this is unlikely to invalidate the result. The trick is to avoid computational algorithms that allow small errors to blossom into big ones.

Quantum error correction is one of the most active fields in quantum computation. It's really an engineering problem: a matter of good design of quantum circuits. There's no universal solution, nor is it necessarily the case that an error-correction method that works for small numbers of qubits will remain effective for larger circuits. The solutions are part of the hard graft of turning the theory of quantum computation into real devices. The essential problem of why errors are difficult to manage, however, goes to the roots of how quantum mechanics works.

•

While quantum no-cloning is a headache for the quantum computer engineer, it also offers technological opportunities of its own. Inspired by a much earlier suggestion of physicist Stephen Wiesner of Columbia University, Charles Bennett and Gilles Brassard pointed out in the early 1980s that quantum correlations between entangled states can be exploited to send information encoded in qubits that is immune to eavesdropping. The signal could never be intercepted and read without the tampering being detected. This was the start of the technology known as quantum cryptography.

The idea goes like this. Alice encodes her message in two entangled qubits – 1s and 0s represented by, say, the polarization states of pairs of photons. One photon from each entangled pair is then sent to Bob, who measures their states and so decodes the message.

There are several methods (protocols) for making this process tamper-proof, but in general they exploit the fact that if an eavesdropper (Eve) intercepts any of these photons, with an unknown polarization state, she can't copy it because she doesn't know exactly how Alice prepared it and so is snookered by quantum no-cloning. She can measure the polarization state, but in doing so she destroys that state and can't make a perfect replica to send on to Bob, hiding her intervention.

In the protocol initially proposed by Bennett and Brassard in 1984, Alice prepares the entangled photons in two different ways. Only if Bob measures his photons using the same scheme that Alice used to prepare it will he get the correct result; otherwise he gets a 1 or 0 at random. He has to guess, and so by chance will only use the right measurement scheme half the time. But even if he chooses the wrong measurement method, half of those 'incorrectly measured' photons will still, by chance, be correct. So 75% of the data will be right. Then if Bob tells Alice, using an ordinary, insecure 'classical' communication channel, which measurement method he used for each photon, Alice can check her records and tell him which 25% to discard. The rest should match perfectly – which they can check by comparing records for just a small subset of the data, again using the classical channel.

If, however, Eve intercepts some or all photons, measures them and generates new ones to send on to Bob, she'll only guess the correct way to prepare them half of the time, by chance. The upshot is that, once Bob has discarded the one in four results he measured wrongly, he and Alice then find there is still not a perfect match in their 'checking' subset. This tells them that the transmission was intercepted.

So it's not that the optical signal can't be intercepted; it can be. But Eve's snooping *can never be hidden* from Alice and Bob. Quantum cryptography, says Brassard, 'offers an unbreakable method for code-makers to win the battle once and for all against any possible attack available to codebreakers'.

Bennett, Brassard and their students achieved a crude experimental demonstration of their protocol in 1989: adequate to show that the idea could be made to work but nowhere near good enough to be practically useful. Since then the technology has been refined to the point that there are now private companies selling quantum-cryptographic devices, such as ID Quantique in Switzerland. Such technology was used – more as a proof of principle than as an essential precaution – to encrypt and transmit the results of the Swiss federal election in 2007 from a data entry centre to the government repository in Geneva. Installation has begun on a Chinese fibre-optic quantum communications network stretching from Shanghai to Beijing for secure transmission of government and financial data.

All the same, implementations of quantum cryptography aren't yet perfect, and the flaws provide loopholes for 'quantum hackers'. This isn't the disreputable and piratical business it might sound, for the point of such attacks is not (yet?) to gain illicit access to sensitive real-world information. Rather, it's to probe the limits of the theory: to figure out what is and is not made possible by the laws of quantum mechanics, and thereby hopefully gain a deeper understanding of the theory itself.

•

The no-cloning rule prohibits us from making an exact copy of an unknown or arbitrary quantum state. But if you're clever about it, you can *transfer* unknown information in the quantum state of one particle into that of another, if the two are entangled. The second particle then becomes a replica, but in the process the information is necessarily erased in the first.

To all intents and purposes, it then looks as if the first particle has vanished from its original location and reappeared elsewhere. It hasn't really performed any dematerialization; but if the replica is genuinely indistinguishable from the original, the result is the same. That's why, when this possibility was first recognized in 1993 by Asher Peres and Bill Wootters, who proposed to call it 'telepheresis' (loosely, 'long-distance manifestation'), Charles Bennett suggested a more catchy name: quantum teleportation.

Like quantum cryptography, the 'teleportation' procedure involves a sharing of two entangled particles A and B between the sender (Alice) and receiver (Bob) – the entanglement sets up what is often called a quantum channel, although it's misleading to think that anything is 'sent', spooky-action style, along it. Alice also has a particle C, whose state she may or may not know (either option works) and which she aims to teleport into the state of Bob's particle B.

To do that, Alice makes a particular kind of simultaneous measurement on her particles A and C which is in effect of the same kind as that in a Bell test (page 174). This doesn't reveal what state C is in; but because of the entanglement between A and B, it places B in a state that can be turned into whatever state C originally had, *if* Bob applies the right operation to it. By making her measurement, Alice has erased that state from C itself, so the original and its copy never coexist.

What is the operation Bob needs to apply to B to complete the teleportation? He can deduce that from the outcome of Alice's Bell measurement, which she has to communicate to him by some classical means. Once he gets that information (which can reach him no faster than the speed of light), he can transform B into a replica of C.

Is this teleportation in any real sense? Importing ideas into science from myth, fantasy and science fiction is always a mixed blessing. It helps people to grasp the central concept, but also risks raising unrealistic or inappropriate expectations. When experimental quantum teleportation was first reported (for photons) by Anton Zeilinger's group in Vienna in 1997, the newspapers were full of speculations about *Star Trek*-style devices that will send us instantly to the other side of the world. It's not even clear what quantum teleportation could possibly mean in such a context, any more than we can think about describing a mental state, or a cat, in a wavefunction. Quantum teleportation could, in principle, be a neat way of moving information around in a quantum computer or a data network. But when newspaper stories tell you that using it as a handy means of human travel is 'still a long way off', what they mean is that the fantasy has become confused with the reality. Which is, in quantum mechanics, an occupational hazard.

Quantum computers don't

necessarily perform ‘many calculations at once’

I'll say it upfront: no one fully understands how quantum computers work.

Yes – we can calculate and predict what a quantum device should do without being too clear about how it does it.

You wouldn't guess this from most popular accounts, and indeed from some technical ones too. We are usually told that quantum computers are faster than classical computers because, by encoding information in superpositions of qubits, they can perform many calculations at once, generating all the possible answers. Then the collective wavefunction of the qubits is collapsed in some clever fashion that guides it into precisely that state which corresponds to the correct or optimal answer.

It sounds attractively plausible. But that's probably not the way such computers will work in general, and perhaps it's not how they work at all.

•

The notion of 'quantum parallelism' comes from the seminal early work in quantum computing in the 1980s by David Deutsch. Researchers in the field know that it is probably a misleading explanation, but some will admit to using it for convenience, especially when speaking to non-specialists and journalists. Others are more forthright about the shortcomings of 'quantum parallelism' and believe that the source of the speed-up has an entirely different character.

If that's not really how a quantum computer gets its speed, then what is? There's no consensus, but this ignorance should be embraced rather than hidden or patched over with half-truths. For, as we've seen, quantum computing is bound up with some of the big foundational questions of the field. Just as we can use quantum theory to correctly predict the outcomes of experiments on double-slit diffraction or Bell-test entanglement yet without being able to say exactly why, so it is clear that quantum computing works in principle but we can't say exactly why. And the questions are of the same kind.

Deutsch's original formulation of quantum computing reflects his deep commitment to the Many Worlds Interpretation of quantum mechanics, which holds that every possible state of a wavefunction corresponds to a physical reality – an idea examined in the next section. In this view, a quantum computer actually does its work *in many worlds* at once, whereas a classical computer has only one world in which to work. Deutsch was convinced that the very possibility of quantum computing lends support to the Many Worlds hypothesis.

But many quantum-computation theorists suspect that the real key to quantum speed-up is not parallelism (let alone parallel computing in Many Worlds) but entanglement. The computation uses the entangled relationships between qubits to manipulate them all together, without having to do many repetitive operations on each qubit individually. That can cut out a lot of bother, because it means that you can leap between many-qubit states without having to work through intermediate steps that would have been taken on a classical computer. Entanglement means that computational steps somehow 'count for more' on a quantum computer. Thanks to quantum non-locality, by making an intervention *here* you seem able to influence what goes on *there*. And so by doing one thing to the qubits, you get a whole lot more for free.

It's not the only way of looking at the situation, though. Others feel that quantum-computational speed-up is more about the *interference* that is possible between quantum states: the fact that the probability of two quantum states is not the same as the sum of their individual probabilities. Admittedly, entanglement is itself a manifestation of interference, because it sets up correlations between the individual states. But it's possible to have interference without entanglement, as in the double-slit experiment, say.

And indeed it's now clear that entanglement, although a requirement for most quantum-computing schemes, is not an *essential* resource. Maarten van den Nest of the Max Planck Institute for Quantum Optics in Garching has outlined a theoretical method of quantum computation that will work as well as one that demands entanglement yet with only a vanishingly small amount of it. (Why then not say 'with none'? It's because van den Nest's method starts with some entanglement and shows that this can be gradually reduced, as close as you like to zero, without degrading the performance.) Entanglement might, then, play no decisive role for quantum speed-up. Certainly, the converse is true: according to theory, high amounts of entanglement – or for that matter, simply of quantum interference – in a quantum computer do not guarantee that it will be faster than a classical machine.*

If it's not from the vast variety of possible qubit states and the availability of other worlds, nor from entanglement or interference either, that the speed of quantum computers derives, then from what?

Another candidate source is contextuality (page 190): the dependence of quantum outcomes on the context of measurement. Joseph Emerson of the University of Waterloo in Canada and his colleagues have argued that this is the hidden resource needed for at least some forms of quantum speed-up. But the debate goes on.

•

We don't know, either, how deeply we can draw from the quantum well. An additional resource for making computation and communication *even more* efficient, beyond what entangled qubits can provide, has been identified by Lucien Hardy, working at the Perimeter Institute in Waterloo, Canada, and independently by Giulio Chiribella and his co-workers at the University of Pavia in Italy. It involves the counter-intuitive manoeuvre of creating a superposition in the *direction* in which information is passed between a sending and receiving quantum logic gate – so that it isn't possible to say which is which.

In normal computer circuits, and indeed those of conventional quantum computers, information is shunted

* In fact it turns out that a quantum computer can have *too much* entanglement, in the sense that above a certain threshold it becomes possible to imitate the performance of the device using a classical computer: the quantum advantage is lost.

from one device to another: a logic device receives 1s and 0s from one place, say, manipulates them, and passes them on. But in a scheme proposed by the Pavia team, a qubit acts as a switch to control the direction in which the signal is passed between two such devices, which we might denote box A and box B. As it is a qubit, it can be placed in a superposition – meaning that we can think (albeit not too literally, as we now know) of the information simultaneously moving from box A to box B and in the opposite direction.

Well, perhaps that's not so odd at first sight? Things can travel two ways at once, after all. If we imagine a channel connecting a box filled with one gas to a box filled with another, we can easily imagine each gas simultaneously diffusing down the channel to the other box, in opposite directions. But that's not what we're talking about here. The information is not like a gas, but could be a single bit – like a ball. So here the ball looks as though it is passing in both directions at once.

What is particularly perplexing about this situation is that it seems to leave indeterminate the direction of *causality*: is box A acting on box B, or vice versa? We can't meaningfully say.

Researchers in Vienna have created these causal superpositions in experiments with photons. And they have shown that for certain types of computation, a quantum switch that permits such states to be made can simplify the computational process: the number of qubits that must be exchanged between gates to carry out the task may be considerably less than it is if the units are merely entangled. A quantum computer that scrambles causation this way becomes faster still.

•

The fundamental 'how' of quantum computation doesn't, in truth, matter very much to many researchers trying to make quantum computers that work. Their concerns are immediate practical issues of the engineering, such as how to achieve longer coherence times for entangled qubits, how to carry out as many computational operations as possible within that window of coherence, how to couple and uncouple qubits controllably, and so on.

Some say that the notion of a 'resource' that enables quantum computing is in any case misleading. The idea that there is some quintessential quantum spice that will one day be bought by the ounce in computer stores much as today we pay for gigabytes of memory is, they insist, a fiction.

All the same, it seems that there is at least one criterion for whether a computation can *only* be achieved using quantum shortcuts. This is supplied by the concept of quantum discord: the measure of 'quantumness' introduced by Wojciech Zurek (page 233). It has been shown that if a particular computational process is 'discord-free', then it can surely be carried out efficiently on a classical computer instead. 'This amounts to saying', Zurek explains, 'that whatever the magic quantum ingredient is, it resides in states that truly have "quantumness"' – in other words, in states correlated so that the mutual information they share possesses some quantum discord.

Still, there may not in fact be a one-size-fits-all answer to how quantum computers work. The particular 'ingredient' needed to release the power of the quantum approach might be different for different implementations (even if each of them incorporates some quantum discord). Then any simple account of the process is doomed to be incomplete, if not actively misleading.

Understanding exactly how quantum mechanics can improve computing may turn out to provide insights into one of the deepest questions of the field: what quantum information really is and how it can be transmitted and altered. This isn't a theoretical issue divorced from the realities of making devices. We've seen that so far only a small number of algorithms have been proposed that are well suited to particular problems, such as factorization and searching. There isn't a straightforward way of making use of what quantum mechanics has to offer, and designing good quantum algorithms is a very difficult task. That task should be easier if we had a better grasp of which aspect of quantum mechanics supplies the potential advantages.

But it's not clear if we will ever really grasp that. 'My own feeling on this issue is that the quantum speed-up is a property of quantum mechanics as a whole and is not something you can definitively pinpoint the source of', says the mathematician Daniel Gottesman. 'If you have "enough" of quantum mechanics available, in some sense, then you have a speed-up, and if not, you don't.'

There's a Bohr-like appeal to this view – a vision of quantum mechanics as a kind of Thing in Itself, irreducible to a more fundamental or fragmented description. How do quantum computers work? By using quantum mechanics.

If that seems an unsatisfying and unedifying answer, rest assured it's for the same reasons that many people (justifiably) feel this way about the Copenhagen Interpretation. Yet such ambiguity may at least have the virtue that it leaves space for researchers to draw inspiration from the diversity of views about what quantum mechanics 'means'. After all, even if a quantum computer does indeed require only one universe, David Deutsch's vision of a multiplicity of worlds helped him to launch the field. Critics might dismiss his view while embracing what it produced. It's a reminder that in science – and this is arguably truer than ever in a contested field like quantum mechanics – it's as worthwhile for an idea to be productive as it is for it to be 'right'.

Deutsch's Many Worlds convictions can, at least in this regard, therefore be considered productive. But what are the chances that they are right?

There is no other

‘quantum’ you

If Murray Gell-Mann was right that Niels Bohr brainwashed a generation of physicists to accept the Copenhagen Interpretation, either his influence has waned or he didn't do a very good job in the first place.

For in an informal poll conducted at an international meeting in 2011 on 'quantum physics and the nature of reality', fewer than half of the attendees professed an allegiance to Bohr's position. True, his was still the most popular interpretation by a considerable margin. But it can hardly claim to represent a consensus.

Sixteen years previously, another show of hands had been taken by the MIT physicist Max Tegmark at a similar meeting in Maryland. The Copenhagen Interpretation triumphed on that occasion too, albeit also without a majority. But Tegmark was delighted to note that in second place was his own favoured view of quantum mechanics: the Many Worlds Interpretation (MWI).*

You might well have heard of it already, for several popularizers of quantum theory have described it at some length and campaigned on its behalf. It is the most extraordinary, alluring and thought-provoking of all the ways in which quantum mechanics has been interpreted. In its most familiar guise, it suggests that we live in a near-infinity of universes, all superimposed in the same physical space but mutually isolated and evolving independently. In many of these universes

* It's been suggested, however, that what these polls really tell you is who organized the meeting.

there exist replicas of you and me, all but indistinguishable yet leading other lives.

The MWI illustrates just how peculiarly quantum theory forces us to think. It is an intensely controversial view. Arguments about the interpretation of quantum mechanics are noted for their passion, as disagreements that can't be settled by objective evidence are wont to be. But when the MWI is in the picture, those passions can become so extreme that we must suspect a great deal more invested in the matter than simply the resolution of a scientific puzzle.

The MWI is qualitatively different from the other interpretations of quantum mechanics, although that's rarely recognized (or admitted), and it's why I have postponed considering it until now. For the interpretation speaks not just to quantum mechanics itself but to what we consider knowledge and understanding to mean in science. It asks us what sort of theory, in the end, we will demand or accept as a claim to *know* the world.

•

After Bohr articulated and refined what became known as the Copenhagen Interpretation in the 1930s and 40s, it seemed that the central problem for quantum mechanics was the mysterious rupture created by observation or measurement, which was packaged up into the rubric of 'collapse of the wavefunction'.

The Schrödinger equation defines and embraces all possible observable states of a quantum system. Before the wavefunction collapses (whatever that means) there is no reason to attribute any greater a degree of reality to any of these possible states than to any other. For remember that quantum mechanics does not imply that the quantum system is actually in one or other of these states but we don't know which. We can confidently say that it is not in *any one* of these states, but is properly described by the wavefunction itself, which in some sense 'permits' them all as observational outcomes. Where then do they all go, bar one, when the wavefunction collapses?

At first glance, the MWI looks like a delightfully simple answer to that mysterious vanishing act. It says that none of the states vanishes at all, except to our perception. It says, in essence, *let's just do away with wavefunction collapse altogether.*

This solution was proposed by the young physicist Hugh Everett III in his 1957 doctoral thesis at Princeton, where he was supervised by John Wheeler. It purported to solve the 'measurement problem' using only what we know already: that *quantum mechanics works.*

But Bohr and colleagues didn't bring wavefunction collapse into the picture just to make things difficult. They did it because *that's what seems to happen.* When we make a measurement, we really do get just one result out of the many that quantum mechanics offers. Wavefunction collapse seemed to be demanded in order to connect the theory to reality.

So what Everett was saying was that this isn't, after all, what reality is. It's not quantum mechanics that is at fault, but our concept of reality. We only *think* that there's a single outcome of a measurement. But in fact all of them occur. We only see one of those realities, but the others have a separate physical existence too.

In effect this implies that the entire universe is described by a gigantic wavefunction: as Everett called it in his thesis, the 'universal wavefunction'. This begins as a superposition of all possible states of its constituent particles – it contains within it all possible realities. As it evolves, some of these superpositions break down, making certain realities distinct and isolated from one another. In this sense, worlds are not exactly 'created' by measurements; they are just separated. This is why we shouldn't, strictly speaking, talk of the 'splitting' of worlds (even though Everett did), as though two have been produced from one. Rather, we should speak of the unravelling of two realities that were previously just possible futures of a single reality.

When Everett presented his thesis (and at the same time published the idea in a respected physics journal), it was largely ignored. It wasn't until 1970 that people began to take notice, after an exposition on the idea was presented in the widely read magazine *Physics Today* by the American physicist Bryce DeWitt.

This scrutiny forced the question that Everett's thesis had somewhat skated over. If all the possible outcomes of a quantum measurement have a real existence, where are they, and why do we see (or think we see) only one? This is where the many worlds come in. DeWitt argued that the alternative outcomes of the measurement must exist in a parallel reality: another world. You measure the path of an electron, and in this world it seems to go this way, but in another world it went that way.

That requires a parallel, identical apparatus for the electron to traverse. More, it requires a parallel *you* to observe it – for only through the act of measurement does the superposition of states seem to 'collapse'. Once begun, this process of duplication seems to have no end: you have to erect an entire parallel universe around that one electron, identical in all respects except where the electron went. You avoid the complication of wavefunction collapse, but at the expense of making another universe. The theory doesn't exactly *predict* the other universe in the way that scientific theories usually make predictions. It's just a deduction from the hypothesis that the other electron path is real too.

This picture gets really extravagant when you appreciate what a measurement is. In one view, any interaction between one quantum entity and another – a photon of light bouncing off an atom – can produce alternative outcomes, and so demands parallel universes. As DeWitt put it, 'every quantum transition taking place on every star, in every galaxy, in every remote corner of the universe is splitting our local world on earth into myriads of copies'. In this 'multiverse', says Tegmark, 'all possible states exist at every instant' – meaning, at least in the popular view, that *everything that is physically possible* is (or will be) realized in one of the parallel universes.

In particular, after a measurement takes place there are two (or more) versions of the observer where before there was one. 'The act of making a decision', says Tegmark – a decision here counting as a measurement, generating a particular outcome from the various possibilities – 'causes a person to split into multiple copies.' Both copies are in some sense versions of the initial observer, and both of them experience a unique, smoothly changing reality which they are convinced is the 'real world'. At first these observers are identical in all respects except that one observed *this* path (or an *up* spin, or whatever is being measured) and the other *that* path (or a *down* spin ...). But after that, who can say? Their universes go their separate ways, launched on a trajectory of continual unravelling.

You can probably see why the MWI is the interpretation* of quantum mechanics that wins all the glamour and publicity. It tells us that we have multiple selves, living other lives in other universes, quite possibly doing all the things that we dream of but will never achieve (or never dare to attempt). There is no path not taken. For every tragedy, like Gwyneth Paltrow's character being hit by a van in the Many Worlds-inspired 1998 movie *Sliding Doors*, there is salvation and triumph.

Who could resist that idea?

•

There are, of course, some questions to be asked.

For starters, about this business of bifurcating worlds. No 'splitting' is implied by the Schrödinger equation itself: it tells us only that quantum systems evolve in a unitary way, so that superpositions remain superpositions and different states stay different. How, then, does a split happen?

That is now seen to hinge on the issue of how a microscopic quantum event gives rise to macroscopic, classical behaviour through decoherence. Parallel quantum worlds have split once they have decohered, for by definition decohered wavefunctions can have no direct, causal influence on one another. For this reason, the theory

* Again, be warned: there are several variants of the Many Worlds Interpretation, so it is sometimes hard to make statements that apply to them all.

of decoherence developed in the 1970s and 80s helped to revitalize the MWI by supplying a clear rationale for what previously seemed a rather vague contingency.

In this view, splitting is not an abrupt event. It evolves through decoherence, and is only complete when decoherence has removed all possibility of interference between universes. While it's popular to regard the appearance of distinct worlds as akin to the bifurcation of futures in Luis Jorge Borges' story 'The Garden of Forking Paths', a better analogy might therefore be something like the gradual separation of shaken salad dressing into layers of oil and vinegar. It's then meaningless to ask *how many* worlds there are – as the philosopher of physics David Wallace aptly puts it, the question is rather like asking 'How many experiences did you have yesterday?' You can identify some of them but you can't enumerate them.

What we *can* say a little more precisely is what kind of phenomenon causes splitting. In short, it must happen with dizzying profusion. Just within our own bodies, few if any biomolecular interactions (such as protein molecules encountering each other in cells) can be expected to produce long-lived superpositions. There would then be at least as many splitting events affecting each of us every second as there are encounters between our molecules in the same space of time. Those numbers are astronomical.

The main *scientific* attraction of the MWI is that it requires no changes or additions to the standard mathematical representation of quantum mechanics. There is no mysterious, ad hoc and *non-unitary* collapse of the wavefunction. And virtually by definition it predicts experimental outcomes that are fully consistent with what we observe.

But if we take what it says seriously, it soon becomes clear that the conceptual and metaphysical problems with quantum mechanics aren't banished by virtue of this apparent parsimony of assumptions and consistency of predictions. Far from it.

•

The MWI is surely the most polarizing of interpretations. Some physicists consider it almost self-evidently absurd; 'Everettians', meanwhile, are often unshakeable in their conviction that this is the most logical, consistent way to think about quantum mechanics. Some of them insist that it is the *only* plausible interpretation of quantum mechanics – for the arch-Everettian David Deutsch, it is not in fact an 'interpretation' of quantum theory at all, any more than dinosaurs are an 'interpretation' of the fossil record. It is simply what quantum mechanics *is*. 'The only astonishing thing is that that's still controversial', Deutsch says.

My own view is that the problems with the MWI are overwhelming – not because they show it must be wrong, but because they render it incoherent. It simply cannot be articulated meaningfully. These objections need to be considered in detail to make their weight fully apparent, but I'll attempt to summarize them.

First, let's dispense with a *wrong* objection. Some criticize the MWI on aesthetic grounds: people object to all those countless other universes, multiplying by the trillion every nanosecond, because it just doesn't seem *proper*. Other copies of me? Other world histories? Worlds where I never existed? Honestly, whatever next! This objection is rightly dismissed by saying that an affront to one's sense of propriety is no grounds for rejecting a theory. Who are we to say how the world *should* behave?

A stronger objection to the proliferation of worlds is not so much all this extra stuff you're making, but the insouciance with which it is made. Roland Omnès says the idea that every little quantum 'measurement' spawns a world 'gives an undue importance to the little differences generated by quantum events, as if each of them were vital to the universe'. This, he says, is contrary to what we generally learn from physics: that most of the fine details make no difference at all to what happens at larger scales.

But one of the most serious difficulties with the MWI is what it does to the notion of self. What can it mean to say that splittings generate copies of me? In what sense are those other copies 'me'?*

Brian Greene, a well-known physics popularizer with Everettian inclinations, insists simply that 'each copy *is* you'. You just need to broaden your mind beyond your parochial idea of what 'you' means. Each of these individuals has its own consciousness, and so each believes he or she is 'you' – but the real 'you' is their sum total.

There's an enticing frisson to this idea. But in fact the very *familiarity* of the centuries-old doppelgänger trope prepares us to accept it rather casually, and as a result the level of the discourse about our alleged replica selves is often shockingly shallow – as if all we need

* Some say that questions about identity in the MWI should be shelved simply because they are 'not physics': a craven retreat into a disciplinary shell that is akin to a factory manager refusing to take any responsibility for the noxious effluent once it leaks beyond the boundaries of his premises.

contemplate is something like teleportation gone awry in an episode of *Star Trek*. We are not being astonished, but rather, flattered by these images. They sound transgressively exciting while being easily recognizable as plot lines from novels and movies.

Tegmark waxes lyrical about his copies: 'I feel a strong kinship with parallel Maxes, even though I never get to meet them. They share my values, my feelings, my memories – they're closer to me than brothers.' But this romantic picture has, in truth, rather little to do with the realities of the MWI. The 'quantum brothers' are an infinitesimally small sample cherry-picked for congruence with our popular fantasies. What about all those 'copies' differing in details graduating from the trivial to the utterly transformative?

The physicist Lev Vaidman has thought rather carefully about this matter of quantum youness. 'At the present moment there are many different "Levs" in different worlds', he says, 'but it is meaningless to say that now there is another "I". There are, in other words, beings identical to me (at the time of splitting) in these other worlds, and all of us came from the same source – which is "me" right now.'

The 'I' at each moment of time, he says, is defined by a complete classical description of the state of his body and brain. But such an 'I' could never be conscious of its existence. Consciousness relies on *experience*, and experience is not an instantaneous property: it takes time, not least because the brain's neurons themselves take a few milliseconds to fire. You can't 'locate' consciousness in a universe that is frantically splitting countless times every nanosecond, any more than you can fit a summer into a day.

One might reply that this doesn't matter, so long as there's a perception of continuity threading through all those splittings. But in what can that perception reside, if not in a conscious entity?

And if consciousness – or mind, call it what you will – *were* somehow able to snake along just one path in the quantum multiverse, then we'd have to regard it as some non-physical entity immune to the laws of (quantum) physics. For how can it do that when, according to the Schrödinger equation, *nothing else does*?

David Wallace, one of the most ingenious Everettians, has argued that purely in linguistic terms the notion of 'I' can *only* make sense if identity/consciousness/mind is confined to a single branch of the quantum multiverse. Since it is not clear how that can possibly happen, Wallace might then have inadvertently demonstrated that the MWI is *not* after all proposing a conceit of 'multiple selves'. On the contrary, it is dismantling the whole notion of selfhood. It is denying any real meaning of 'you'.

I shouldn't wish anyone to think that I feel affronted by this. But if the MWI sacrifices the possibility of thinking meaningfully about selfhood, we should at least acknowledge that this is so, and not paper over it with images of 'quantum brothers and sisters'.

•

The science-fiction vision of a 'duplicated quantum self' has nevertheless delivered some fanciful, and undeniably entertaining, images. If splitting can be guaranteed by any experiment in which the outcome of a quantum process is measured, then one can imagine making a 'quantum splitter': a handheld device in which, say, an atomic spin is measured and the result is converted to a macroscopic arrow pointing on a dial to 'Up' or 'Down' – which ensures that the initial superposition of spin states is fully decohered into a classical outcome. You can make these measurements as often as you like just by pushing the button on the device. Each time you do (so the story goes), two distinct 'yous' come into being.

What can you do with this power to generate worlds and selves? You could become a billionaire by playing quantum Russian roulette. Your quantum splitter is activated while you sleep, and if the dial says Up then you're given a billion dollars when you wake. If it shows Down then you are put to death painlessly in your sleep. Few people, I think, would accept these odds on a coin toss. But a committed Everettian should have no hesitation about doing so using the quantum splitter. (It's not clear, actually, whether a simple coin toss won't itself act as a splitter.) For you can be certain, in this view, that you'll wake up to be presented with the cash. Of course, only one of 'you' wakes up at all; the others have been killed. But those other yous knew nothing of their demise. Sure, you might worry about the grief afflicted on family and friends in those other worlds. But that aside, the rational choice is to play the game. What could possibly go wrong?

•

You're not going to play? OK, I see why. You're worried about the fact that you're going to die as a result, with absolute certainty. But look, you're going to live and become rich with absolute certainty too.

Are you having trouble comprehending what that means? Of course you are. It has no meaning in any normal sense of the world. The claim is, in words aptly coined by the physicist Sean Carroll in another context (ironically Carroll is one of the most vocal Everettians), 'cognitively unstable'.

Some Everettians have tried to articulate a meaning nonetheless. They argue that, despite the certainty of all outcomes, it is rational for any observer to consider the subjective probability for a particular outcome to be proportional to the amplitude of that world's wavefunction – or what Vaidman calls the 'measure of existence' of that world.

It's a misleading term, since there's no sense in which any of the many worlds 'exists less'. For the 'self' that ends up in any given world, that's all there is – for better or worse. Still, Vaidman insists that we ought rationally to 'care' about a post-splitting world in proportion to this measure of existence. On this basis, he feels that playing quantum Russian roulette again and again (or even once, if there's a very low measure of existence for the 'good' outcome) should be seen as a bad idea, regardless of the morality, 'because the measure of existence of worlds with Lev dead will be much larger than the measure of existence of the worlds with a rich and alive Lev'.

What this boils down to is the interpretation of probabilities in the MWI. If all outcomes occur with 100% probability, where does that leave the probabilistic character of quantum mechanics? And how can two (or for that matter, a thousand) mutually exclusive outcomes all have 100% probability?

There is a huge and unresolved literature on this question, and some researchers see it as the issue on which the idea stands or falls. But much of the discussion assumes,

I think wrongly, that the matter is independent of questions about the notion of selfhood.

Attempts to explain the *appearance* of probability within the MWI come down to saying that quantum probabilities are just what quantum mechanics looks like when *consciousness is restricted to only one world*. As we saw, there is in fact no meaningful way to explain or justify such a restriction. But let's accept for now – just to see where it leads – the popular view of the MWI that two copies of an observer emerge from the one who exists before a measurement, and that both copies experience themselves as unique.

Imagine that our observer, Alice, is playing a quantum version of a simple coin-toss gambling game – nothing as drastic or emotive as quantum Russian roulette – that hinges on measurement of the spin state of an atom prepared in a 50:50 superposition of *up* and *down*. If the measurement elicits *up*, she doubles her money. If it's *down*, she loses it all.

If the MWI is correct, the game seems pointless – for Alice will, *with certainty*, both win *and* lose. And there's no point her saying 'Yes, but which world will *I* end up in?' Both of the two Alices that exist once the measurement is made are in some sense present in the 'her' before the toss.

But now let's do the sleeping trick. Alice is put to sleep before the measurement is made, knowing she will be wheeled into one of two identical rooms depending on the outcome. Both rooms contain a chest – but inside one is twice her stake, while the other is empty. When she wakes, she has no way of telling, without opening the chest, whether it contains the winning money. But she can then meaningfully say that there is a 50% probability that it does. What's more, she can say *before the experiment* that, when she awakes, these will be the odds deduced by her awakened self as she contemplates the still-closed chest. Isn't *that* a meaningful concept of probability?

The notion here is that quantum events that occur for certain in the MWI can still elicit probabilistic beliefs in observers simply because of their ignorance of which branch they are on.

But it won't work. Suppose Alice says, with scrupulous care, 'The experience I will have is that I will wake up in a room containing a chest that has a 50% chance of being filled or empty.' The Everettian would say Alice's statement is correct: it's a rational belief.

But what if Alice were to say, 'The experience I will have is that I will wake up in a room containing a chest that has a 100% chance of being empty'? The Everettian must accept this statement as a true and rational belief too, for the initial 'I' here must apply to both Alices in the future.

In other words, Alice Before can't use quantum mechanics to predict what will happen to her in a way that can be articulated – because there is no logical way to talk about 'her' at any moment except the conscious present (which, in a frantically splitting universe, doesn't exist). Because it is logically impossible to connect the perceptions of Alice Before to Alice After, 'Alice' has disappeared. You can't invoke an 'observer' to make your argument when you have denied pronouns any continuity.

What the MWI really denies is the existence of facts at all. It replaces them with an experience of pseudo-facts (we *think* that this happened, even though that happened too). In so doing, it eliminates any coherent notion of what we can experience, or have experienced, or are experiencing right now. We might reasonably wonder if there is any value – any meaning – in what remains, and whether the sacrifice has been worth it.

•

You might in any case want to ask: where *are* all these 'other worlds' anyway? The usual answer is that they are in Hilbert space – the mathematical construct that contains all the possible solutions of the variables in the Schrödinger equation. But Hilbert space is a construct – a piece of math, not a place. As Asher Peres has put it, 'The simple and obvious truth is that quantum phenomena *do not* occur in a Hilbert space. They occur in a laboratory.' If the Many Worlds are in some sense 'in' Hilbert space, then we are saying that the equations are more 'real' than what we perceive: as Tegmark puts it, 'equations are ultimately more fundamental than words' (an idea curiously resistant to being expressed without words). Belief in the MWI seems to demand that we regard the math of quantum theory as somehow a fabric of reality. We have nowhere to put those Many Worlds, except in our equations. Some physicists suspect that this amounts to falling so deeply in love with your mathematical tools that you decide to live in them.

The issue is whether, like Bohr, you believe that quantum mechanics provides a prescription for evaluating the possible outcomes we might observe when we look at the quantum world, or whether you regard the Schrödinger equation as an inviolable and universal law that describes – in some sense *is* – reality.

But it goes even deeper than that. How we feel about the MWI depends on what we demand from science as a system of knowledge.

Every scientific theory (at least, I cannot think of an exception) is a formulation for explaining why things in the world are the way we perceive them to be. This assumption that a theory must recover our perceived reality is generally so obvious that it is unspoken. The theories of evolution or plate tectonics don't have to include some element that says 'you are here, observing this stuff'; we can take that for granted.

But the MWI refuses to grant it. Sure, it claims to explain why it *looks* as though 'you' are here observing that the electron spin is *up*, not *down*. But actually it is not returning us to this fundamental ground truth at all. Properly conceived, it is saying that there are neither facts nor a you who observes them.

It says that our unique experience as individuals is not simply a bit imperfect, a bit unreliable and fuzzy, but is a complete illusion. If we really pursue that idea, rather than pretending that it gives us quantum siblings, we find ourselves unable to say anything about anything that can be considered a meaningful truth. We are not just suspended in language; we have denied language any agency. The MWI – if taken seriously – is *unthinkable*.

•

If indeed the MWI were a consistent and coherent interpretation of quantum mechanics that allowed for nothing but the unitary evolution of the Schrödinger equation, we would be well advised to take it.

But it is, sadly, not that. Its implications undermine a scientific description of the world far more seriously than do those of any of its rivals. It tells you not to trust empiricism at all: rather than imposing the observer on the scene, it destroys any credible account of what an observer can possibly be. Some Everettians insist that this is not a problem and that you should not be troubled by it. Perhaps you are not, but I am.

Yet I have pushed hard against the MWI not so much to try to demolish it as to show how its flaws, once brought to light, are instructive. Like the Copenhagen Interpretation (which also has profound problems), it should be valued for forcing us to confront some tough philosophical questions.

What quantum theory seems to insist is that at the fundamental level the world cannot supply clear 'yes/no' empirical answers to all the questions that seem at face value as though they should have one. The calm acceptance of that fact by the Copenhagen Interpretation seems to some, and with good reason, to be far too unsatisfactory and complacent. The MWI is an exuberant attempt to rescue the 'yes/no', albeit at the cost of admitting both of them at once. That this results in an *inchoate* view of macroscopic reality suggests we really can't make our macroscopic instincts the arbiter of the situation. And *that*, I would argue, is the value of the Many Worlds: they close off an easy way out. It was worth admitting them in order to discover that they are a dead end. But there is no point then sitting there insisting we have found the way out. We need to go back and keep searching.

Where Copenhagen seems to keep insisting 'no, no and no', the MWI says 'yes, yes and yes'. And in the end, if you say everything is true, you have said nothing.

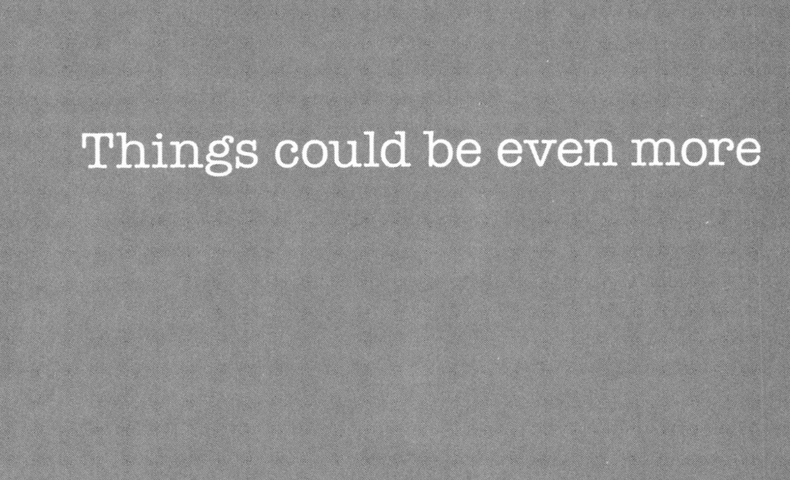
Things could be even more

‘quantum’ than they are
(so why aren’t they)?

I hope you see now why it is time to change the tune of quantum theory. The traditional description in terms of probability waves of particle-like entities is useful for retaining a conceptual link with our familiar classical world of objects moving along paths, and for showing how quantum mechanics differs. It's intuitively helpful to talk about atoms and photons this way. But in the end it leaves us with a strange hybrid theory, peppered with types of behaviour that don't seem to quite make sense: superposition, non-locality, contextuality. We end up having to admit that this isn't really how things are, but just a manner of speaking: an attempt to tell a story despite the inadequacy of our repertoire of narrative devices. When we get down to the fundamentals – to ask what the theory really 'means' – quantum mechanics starts to look like a rather cumbersome, impromptu concoction that we feel compelled to excuse with talk of weirdness.

Increasingly, it looks more logical to frame quantum mechanics as a set of rules about *information*: what is and isn't permissible when it comes to sharing, copying, transmitting and reading it. What distinguishes the quantum world of entanglement and non-locality from the everyday world where such things can't be found is a kind of information-sharing between quantum systems that allows us to find out about one of them by looking at the other. Non-locality is a baffling concept when we think in terms of particles with certain properties located in space, but is perhaps less so when we consider what it means to have knowledge of a quantum system.

Quantum non-locality is the escape clause that rescues quantum mechanics from the 'paradox' Einstein perceived in entanglement – specifically, that it appeared to violate special relativity. Non-locality lets influences *seem* to propagate across space instantaneously while forbidding us from actually sending any meaningful information (indeed, from sending anything at all) that fast. Our intuitive notions of causality – *this* dictating *that* – are salvaged, but only by taking a somewhat broader view of what cause and effect may mean. Einstein's 'spooky action at a distance' vanishes once we think not in terms of pseudo-classical particles interacting via forces but in terms of where in a quantum system information can reside and how it can be probed and correlated.

Yakir Aharonov, who studied under David Bohm, pointed out that this contrivance seems almost fiendishly ingenious: as if quantum mechanics is designed to come as close to violating relativity as it dares, without actually doing so. Could that be a clue to its real nature? Does it show us causation at the brink of breakdown? Or to put it another way, might it be that non-locality is simply in the nature of things, and relativity is the only thing that limits its influence? It's an intriguing idea. If it were true, it could help us to see how these two fundamental theories of the world fit together.

But it turns out that the story doesn't go that way. It is possible to imagine a world that is *even more non-local*, yet still consistent with special relativity – a world that we might call 'super-quantum'.

This became clear in the late 1990s through the work of the physicists Sandu Popescu and Daniel Rohrlich. Their insight offers a new perspective on the question of why quantum mechanics is the way it is. When we encounter entanglement, the invitation is to wonder why the world behaves in this strange non-local way. But Popescu and Rohrlich encourage us to approach the matter from the other direction by asking: 'Why, if things could be even more strange and non-local (without breaking known physical laws), are they not so?'

Understanding why the strength of quantum non-locality is limited in this way could offer a hint about where quantum mechanics comes from in the first place.

•

What do I mean by 'more non-local'? Here's a story about that.

You know Alice and Bob? Of course. Well, each has a black box that dispenses a toy dog or cat when fed with a coin – like something you might find in an amusement arcade. The boxes will accept only dimes or quarters, and which toy is dispensed depends on which coin you put in.

There are rules governing the inputs and outputs:

Rule 1: A dime into Alice's box always produces a cat.
Rule 2: If Alice and Bob both put in quarters, then the boxes will between them spit out one cat and one dog.
Rule 3: *Any* other combination of coins will produce either two cats or two dogs.

Why these particular rules? Think of them as just arbitrary features of how the boxes are designed. (Of course, they're not really arbitrary at all, but are chosen to deliver a specific set of circumstances, as we'll see.)

Can we find combinations of the various inputs and outputs that will satisfy these rules?

If Alice puts in a dime, she'll get a cat (Rule 1). So Bob's box must also produce a cat regardless of which coin he puts in – because a dime in both boxes must produce two cats (Rule 3), while we can only get a cat and a dog if both put in a quarter (Rule 2).

So the boxes can be assigned these input/output relationships:

Alice: dime → cat
Bob: dime → cat
Bob: quarter → cat

The only remaining option to specify is what happens when Alice puts a quarter into her box. If *both* of them put in quarters then we have to get a cat and dog (Rule 2). So a quarter in Alice's box must produce a dog (because Bob's quarter gives a cat):

Alice: quarter → dog

But this won't do. It means that if Alice puts in a quarter and Bob puts in a dime, we get a dog and a cat. Yet we're only meant to get that if both put in quarters (Rules 2 and 3). So that particular combination doesn't work – it breaks the rules.

Can we find some other arrangement of inputs and outputs that does better than this? No we can't (try it yourself). No matter how you juggle it, you'll find that you can only satisfy the rules in three out of the four possible cases: a maximum success rate of 75%.

The rules for Alice's and Bob's boxes will inevitably be violated at least one time in four.

What, though, if Alice's and Bob's boxes can switch their output depending on what the other did? Imagine some communication link between them so that, say, a dime inserted into Bob's box can produce either a cat or a dog, depending on what Alice put in hers. Then you can do better.

Let's say we run a wire between Alice and Bob's boxes which can send an electrical signal between them – so that what Alice chooses to put into hers affects what comes out of Bob's. If Alice puts in a dime (giving a cat), then a dime in Bob's box will produce a cat. But if Alice puts in a quarter (giving a dog) then Bob's dime will produce a dog. That's now consistent with Rule 3.

Alice and Bob's boxes are wired together so that they can communicate: the input to one can affect the output to the other.

Complete success! The only drawback is that this trick can't work instantaneously, because, as special relativity insists, no signal can pass down the wire from Alice's box to Bob's faster than the speed of light. Sure, light travels very fast, but all the same it takes a finite time to get anywhere. If Alice is in Edinburgh and Bob is in Fiji, then Bob will have to wait for a little more than a tenth of a second for Alice's signal to reach him before he puts his coin in. Relativity denies us 100% success *instantaneously*.

But what if these two boxes were governed by the laws of quantum mechanics, so that they can become entangled and 'communicate' not via an electrical connection but using quantum non-locality? Now Bob's box can instantaneously 'use' some information about what Alice's box has done to switch its output. It's possible to calculate by just how much quantum communication* improves the instantaneous success rate. And we find that it doesn't permit success every time, but about 85% of the time – not perfect, but better than the classical boxes.

You might have figured out already that this is a crude analogue of John Bell's experiment to probe entanglement and hidden variables. His scheme for measuring the spins of correlated particles is similar to Alice and Bob looking at (that is, measuring) the binary outputs (cat or dog) of their respective boxes and seeing how they are related. Bell conceived of a particular set of measurements for which classical rules would forbid these correlations from being greater than a certain threshold, whereas with quantum rules of entanglement this threshold can be exceeded. Likewise, the instantaneous success rate for Alice and Bob's boxes is greater for the non-local quantum case than it can be for the classical case, because of the correlations between their outputs.

Is the quantum link between Alice's and Bob's boxes as good as it gets? Or can we imagine a set of instantaneous

* I can't stress it enough: the quantum non-locality involved in entanglement is not *really* a kind of communication across space, although sometimes – if we're careful – that can be a useful metaphor for talking about it.

information-sharing rules that allow Alice and Bob to satisfy the rules *all the time*? Yes, Popescu and Rohrlich showed, we can. We can permit the boxes even more non-local exchange of information than is granted by quantum mechanics, *still without violating relativity*. These super-quantum boxes have become known as Popescu–Rohrlich (PR) boxes.

Their improved performance comes from more efficient sharing of information. In general, communications are very inefficient because they involve exchanging lots of information that doesn't feature in the final answer. This seems to be a fundamental problem for classical information, which is necessarily local: it's fixed in one place. Suppose, say, you and I want to arrange a meeting. We're both very busy, but we compare diaries by phone. We might hit on a suitable date by randomly asking 'Are you free on 6 June?' and so on. But that could take some time if our diaries are very full. To compile a complete list of the days on which we can meet, we have to exchange information about our availability on every single day of the year.

Suppose we look instead for an answer to what sounds like a simpler question: whether the number of possible days we're both free to meet is even or odd. Admittedly that seems a strange thing to ask, since it doesn't exactly help with the original problem of finding a date to meet. But it looks like it should be a simpler thing to decide, because the answer is just *one bit* of information: say, 0 for 'even', 1 for 'odd'.

All the same, we're no better off. The only way we can deduce the value of this single bit is for me again to list every day in the year that I'm free, and you to compare against your calendar. We have to send all those dates just to get a one-bit answer. In fact, *any* problem of comparing data that is recorded classically (written down in a diary, say) can be shown to be equivalent to – and thus as inefficient as – this one.

If we can somehow quantum-entangle our diaries, we don't have to exchange so much information to find the answer to our question: non-locality can reduce some of this redundancy of information sharing. But not all of it.

If we have PR boxes, however, they can remove *all* the redundancy. The question of whether the total number of days on which we could meet is even or odd can be answered by feeding into each of our PR boxes all the details in our respective diaries and having them exchange just *one bit* of information. A bit for a bit: you can't say fairer than that.

For certain types of information processing like this, there is a sharp boundary between what can be done in quantum mechanics and what can be done with super-quantum PR boxes. Just make non-locality a tiny bit stronger than quantum-mechanical, and immediately you're in the super-quantum realm where information exchange is now as efficient as it could possibly be.

So PR boxes tell us that quantum non-locality is a measure of the efficiency with which different systems can appear to communicate and share information. And quantum mechanics is revealed as a particular set of rules within which some outcomes of information sharing and processing are possible, while some (like Alice and Bob achieving 100% success) are not.

•

Given their super-efficiency, PR boxes could be used to do computation even faster than quantum computers. Could they really exist, though? Sure, the world looks quantum-mechanical, not super-quantum; but it also looked classical for a long time, until we figured out how to spot quantum non-locality. Might it be that we've just failed so far to uncover a stronger, PR-box-type non-locality that exists in the real world, and for which quantum non-locality is just an approximation?

We don't know, although it looks unlikely. But even if quantum non-locality is the best we can hope for, hypothetical PR boxes may offer clues about why that is. The question becomes not so much why nature isn't completely classical, but why it's not 'more quantum'. We should then seek answers not by wondering why, say, objects are described by wavefunctions (or what a wavefunction is anyway), but by looking at the more fundamental matter of how information can be shunted about – how efficient communication in nature can possibly be. What is it that apparently limits* quantum non-locality's ability to make information exchange more efficient?

One clue could be a principle called 'information causality' proposed by Marcin Pawlowski of the University of Gdańsk and his colleagues. It's a different way of stating the restrictions on what Bob can find out about what Alice knows. Let's say that Alice has some data: measurements of the spins of entangled quantum

* This limit, which restricts Alice and Bob to an 85% success rate with quantum boxes, is called Tsirelson's bound, after the Russian scientist Boris Tsirelson (or Cirelson) who first identified it.

particles, or observations of which fluffy toys her box has delivered when fed with coins, or free days in her diary. Bob has his own data, and because Alice has sent him some of her data, he can see that the two sets of data are entangled in some way.

Because of those correlations, Bob can use the rest of his own data to figure out more of Alice's. But how much more? The principle of information causality says that it depends on how much information Alice has sent him, in the following way: Bob can't deduce more of Alice's bits (binary spin up/down, or cat/dog, or days free/not free) than the number of bits she has already sent him.

This doesn't mean Bob can only know what Alice has told him. Rather, the amount Bob can deduce about the data in Alice's set that she *hasn't* already disclosed can be no greater than the amount she *has* disclosed. So if Alice doesn't send him *any* of her data, then Bob can't make guesses about it that are better than random – even if there are quantum correlations between what Alice can see and what Bob can see. That's just a way of saying that anything happening at Alice's end can't instantaneously communicate any information to Bob.

There's a pleasing symmetry to this principle of information causality: it sounds rather as though you can't get out more than you put in. With PR boxes, on the other hand, you *can* – even though you still can't use it to send any information instantaneously. So Pawlowski and colleagues think their postulate of information causality might single out precisely what quantum correlations do and don't permit about information transfer. If so, they argue, then 'information causality might be one of the foundational properties of nature' – in other words, an axiom of quantum theory.

•

Again, all this fits with a growing conviction that quantum mechanics is at root a theory not of tiny particles and waves but of information and its causative influence. It's a theory of how much we can deduce about the world by looking at it, and how that depends on intimate, invisible connections between *here* and *there*.

Let's be clear: the theory of PR boxes is not a fully fledged 'alternative quantum theory'. It's just a sort of 'toy model' that can mimic some of its features. So PR boxes might or might not lead us towards the deep principles of quantum mechanics. Regardless of that, they capture the spirit of the idea that it may be possible and helpful to reframe the theory in a way that ditches the paraphernalia of the past (while allowing it to be recovered as and when needed) and replaces it with a simple set of logical axioms.

And what might *those* look like?

The fundamental laws of

quantum mechanics might
be simpler than we imagine

Go to any meeting about the fundamental principles of quantum mechanics, Chris Fuchs wrote in 2002, 'and it is like being in a holy city in great tumult. You will find all the religions with all their priests pitted in holy war.' Things have not changed a great deal since then.

The problem, Fuchs said, is that all of the priests have the same starting point: the standard textbook accounts of the axioms of quantum theory. Like holy scripture, these documents are ambiguous and obscure. There are several ways to express such axioms, but they tend to go something like this:

1. For every system, there is a complex Hilbert space H.
2. States of the systems correspond to projection operators onto H.
3. Those things that are observable somehow correspond to the eigenprojectors of Hermitian operators.
4. Isolated systems evolve according to the Schrödinger equation.

Even at this late stage in the book I don't expect these axioms to make a great deal of sense to you (although some of the jargon might now seem a little familiar). There are words here I haven't explained, and which I am not going to explain. That's the point: why should we need such obscure terminology in the first place? Where, in this dry linguistic thicket, is the real world?

The mechanical world used to be so much simpler, so much more transparent. In classical physics, nearly all the essentials that you needed were incorporated into Isaac Newton's laws of motion:

1. Every moving object keeps moving at the same speed if no force is applied to it. If it is still to begin with, it stays still.
2. If a force is applied to an object, it accelerates in direct proportion to that force, and in the direction of that force.
3. For every force that one body exerts on another, the other body exerts an equal force back in the opposite direction.

You might not grasp the full meaning of these words either, but I'd wager that you get the gist. These are principles that may be expressed in everyday words, and you don't need to undergo years of specialized study before you can understand something of their content. They relate to, and can be demonstrated by, ordinary experience.

Can it be right that the laws of Newtonian mechanics for our classical world, expressed in concise, prosaic words and sentences, must give way to something with the forbidding, abstract mathematical complexity of the quantum axioms?

Or is this just because *we don't quite know what we're talking about*?

When someone explains something in a complicated way, it's often a sign that they don't really understand it. A popular maxim in science used to be that you can't claim to understand your subject until you can explain it to your grandmother.* (Happily, these days it's acknowledged that grandmothers are as likely as anyone else to be astronomers or molecular biologists, and might find themselves having to explain complex scientific ideas to *you*.)

At the outset I mentioned John Wheeler's proposition that if we really understood the central point of quantum theory, we ought to be able to state it in one simple sentence. That is a matter of faith: there's no guarantee that the world's innermost workings will fit a language developed mostly to conduct trade, courtship and banter. All the same, you can't help suspecting that the complicated, highly technical way in which we are currently forced to express quantum axioms points to a failure to get at what the theory is really about.

Wheeler's conviction is shared by Fuchs, who believes that we will one day tell a story about quantum mechanics – 'literally a story, all in plain words' – that is 'so compelling and so masterful in its imagery that the mathematics of quantum mechanics in all its exact technical detail will fall out as a matter of course'. That story, he says, should not only be crisp and compelling. It should also 'stir the soul'.

It would do for quantum mechanics what Einstein did for classical electromagnetic theory and for ideas about the putative light-bearing ether, that nebulous fabric in

* Here's Richard Feynman again. Tasked with constructing a set of lectures to explain a particularly knotty aspect of physics to freshmen, he finally concluded that he couldn't do it. And he interpreted this, to his eternal credit, not as an indication that the subject was too hard, but that he didn't understand it well enough.

which all physicists believed before Einstein's 1905 paper on special relativity. Here was this untidy assemblage of arcane equations trying to describe what happens to light when measured by a moving observer. It already used several of the tools and concepts that were vital to Einstein, including the striking idea that fast-moving objects seem to shrink in the direction of motion. And the theory kind of worked, but it looked ugly and makeshift.

Then Einstein dispelled the mathematical fog with two simple and intuitive principles:

1. The speed of light is constant.
2. The laws of physics are the same for two observers moving relative to one another.

Grant this much and all else follows. In particular, Einstein's theory implied that, while you could conjure up a rather ad hoc explanation for why the light-bearing ether was not seen in experiments, a more logical and satisfying conclusion was that there simply *is* no such ether. Einstein's clarification didn't just make matters easier; it was also rather exhilarating.

What are the analogous statements for quantum mechanics? To find them, we may have to rebuild quantum theory from scratch: to tear up the work of Bohr, Heisenberg and Schrödinger and start again. This is the project known as *quantum reconstruction.*

The reconstructionists are a diverse bunch of physicists, mathematicians and philosophers. They don't agree on how best to do the rebuilding, but collectively their position has much in common with that of the proverbial Irishman asked for directions to Dublin: 'I wouldn't start from here.'

The programme of reconstruction typically seeks to identify some fundamental quantum axioms – hopefully only a small number of them, and with principles that are physically meaningful and reasonable, and that everyone can agree on. The challenge is then to show that the conventional structure and conceptual apparatus of quantum theory emerge as a corollary of these principles, much as the elliptical orbits of the planets (which looked rather perverse and puzzling in the early seventeenth century) can be shown to emerge from Newton's simple and elegant inverse-square law of gravity.

Why rebuild from scratch, only to end up with what you started with? The quest is driven by a suspicion that what we currently consider to be quantum theory is far more baroque than it needs to be, and that's why it's full of puzzles and paradoxes and arguments about interpretation. 'Despite all the posturing and grimacing over the paradoxes and mysteries, no one asks in any serious way: why do we have this theory in the first place?' says Fuchs. 'They see the task as one of patching a leaking boat, not one of seeking the principle that has kept the boat floating this long. My guess is that if we can understand what has kept the theory afloat, we'll understand that it was never leaky to begin with.'

Maybe, then, the exotic paraphernalia of wavefunctions, superpositions, entangled states and the uncertainty principle are necessary only because we're looking at the theory from the wrong angle, making its shadow odd, spiky, hard to decode. If we could only find the right perspective, all would be clear.

•

According to the way things look right now, the key difference between classical and quantum mechanics is that the first calculates trajectories of objects while the second calculates probabilities (expressed as a wave equation). Its probabilistic character doesn't make quantum mechanics unique in itself: coin tossing is about probability too, but you don't need quantum mechanics to explain it. What makes quantum theory so puzzling is that sometimes what we observe seems to force us to speak as though the quantum coins were both heads and tails at once.

One of the first attempts at quantum reconstruction sought to frame it as a theory about probabilities, albeit with somewhat different rules from those used to understand dice and racehorses. In 2001, Lucien Hardy postulated some probability rules that linked the variables characterizing the state of a system – these could be measures of position, momentum, energy, spin and so on – to the ways we might be able to distinguish the values of these variables by measurement. These rules amount to a way of counting up the probabilities of different experimental outcomes based on some assumptions about how systems can carry information and how they may be combined and interconverted.

Hardy's rules, and what they imply, are simply a generalization and modification of standard probability theory. This model could have been derived in principle as an abstract idea by nineteenth-century mathematicians, before any inkling of the empirical motivations that led Planck and Einstein to launch quantum mechanics. There's no 'quantumness' added to this picture by hand, as it were: we just have a set of hypothetical rules linking the possible states of a system to the possible outcomes of observations of it. One set of rules leads to classical behaviour, but other rules offer greater richness in the link between these two things.

Hardy showed that if we assume the simplest of those elaborations, then out of these postulates arise the basic characteristics of quantum mechanics, such as superposition and entanglement. That's to say, if the outcomes follow these particular probability rules, they look like those we find in states of quantum superposition. It's as if quantum-type behaviour *comes from* a certain kind of probability.

This approach, reformulating quantum mechanics as an abstract 'generalized probability theory' linking inputs (possible states of a system) to outputs (measurement of some property), has since been developed further by Hardy and others, such as Giulio Chiribella, Časlav Brukner, Markus Müller and their coworkers. All have shown how various sets of axioms along similar lines can give rise to characteristically quantum behaviour.

•

Jeffrey Bub of the University of Maryland has followed a similar strategy for building quantum theory – at least, a reduced version of it – from some simple postulates about how information can be encoded in a system and read out from it by observation. Quantum mechanics, says Bub, is 'fundamentally a theory about the representation and manipulation of information, not a theory about the mechanics of nonclassical waves or particles'. That's as clear a statement as you could wish for of why early quantum theory got distracted by the wrong things.*

Sure, says Bub, you can construct quantum mechanics in terms of wavefunctions and quantized states, and you can then weave various interpretational stories around those elements – a Bohmian story, a Bohrian story, an Everettian story. But these are just different ways of reshuffling the same empirical facts. Rather than attempting to 'explain' those experimental observations in terms of underlying principles, we should accept that the observations *define* the principles. That's what Einstein did with special relativity. Observations made by two American scientists in the 1880s showed that the speed of light seems to be constant for all observers, no matter how fast they are moving. Instead of trying to explain it, Einstein accepted that constancy as an axiom and then he figured out the consequences.†

This approach can be framed as what physicists call a 'no-go' principle: a statement that something is simply forbidden. Changing the speed of light in a vacuum is forbidden in special relativity: it is the theory's no-go principle.

* Not everyone agrees, because in quantum mechanics no one ever does. 'Information' was on John Bell's list of 'bad words' which, he wrote, 'however legitimate and necessary in application, have no place in a *formulation* [of quantum theory] with any pretension to physical precision.' His list also included *system, apparatus, environment, observable* and, worst of all, *measurement.*

† It's not clear, however, whether Einstein was at all motivated by, or even knew about, those earlier observations on the speed of light when he proposed special relativity in 1905. He had other reasons to imagine that the speed of light might be constant irrespective of relative motion.

This change of perspective turns on its head the usual relationship between theory and experiment. Typically we make an observation and then think how our existing theories can explain it. But every so often – as with relativity, and indeed with the original quantum mechanics itself – we have to say instead: 'Well then, so this is how things are – doing *that* is simply prohibited. If, then, we start from the assumption that the universe will not permit us to do *that*, what follows?' And what often follows is that we then have to ditch the old ideas and theories and build something new.

So what are the no-go principles of quantum mechanics? Bub argues that these should be principles about *what can be done with information*: how it can be encoded, moved and permuted.

This is really a question about the *logic* that applies to quantum mechanics. It comes down to this. If you describe a system using a kind of algebra in which the various terms in the equations commute – crudely meaning (page 151) that the answers you get don't depend on the order in which you perform the calculations – then what you see is classical behaviour. But if the algebra of your equations doesn't commute – if the order matters – then you get a quantum-type theory.

Remember that this is where the uncertainty principle comes from: the fact that in quantum mechanics some quantities do not commute. Bub believes that non-commutativity is what distinguishes quantum from classical mechanics. This property, he says, is a feature of the way information is fundamentally structured in our universe.

Non-commutativity alone doesn't give you quantum mechanics as we know it, however. It gives you the *possibility* of quantum-like non-local behaviour, but in a whole bunch of quantum-like theories. Bub, together with Rob Clifton and Hans Halvorson, has proposed that we can get closer to the quantum theory we know if we create a non-commutative algebra with just three principles about what can (actually, what cannot – these are no-go principles, remember) be done with information. These are all things that we have found to be true of the existing quantum theory and to be supported by experiments. When we encountered (some of) them earlier, it was rather in the sense of 'Well what do you know, quantum mechanics prohibits X!' But now we need to think of them not as discoveries or outcomes of quantum mechanics, but as axioms. We can then ask: if indeed X is prohibited, what follows?

The three no-go principles of Clifton, Bub and Halvorson are these:

1. You can't transmit information faster than light between two objects by making a measurement on one of them (the condition imposed by special relativity, and called no-signalling).
2. You can't deduce or copy perfectly the information in an unknown quantum state (this is more or less the no-cloning rule).
3. There is no unconditionally secure bit commitment.

Sorry, I sprang that last one on you. It's a bit complicated, but it arises (as the phrase hints) out of considerations connected to quantum communications and cryptography. Bit commitment simply means that one observer in an exchange of information (Alice) sends an encoded bit to the other (Bob). If bit commitment is secure against any subterfuge by Bob, this means that Bob can only decode the bit if Alice supplies some further information about the encoding – namely, the key. And for the exchange to be secure against any kind of cheating by Alice, there must be no way that she can change the value of the bit between sending it and Bob reading it. It isn't possible to guard against all forms of dishonesty from Alice and Bob – if Alice sends Bob a false key, say. But we might at least aspire to so-called *unconditionally secure* bit commitment, which means that, so long as Alice and Bob are honest, the encoding of the bit fixes the value of the bit for sure, and that the key can, as near as dammit (not a technical term), perfectly conceal the encoded information, regardless of whatever technical resources an eavesdropper has at their disposal.

Dominic Mayers of the University of Montreal showed in 1997 that for quantum cryptography this unconditionally secure bit commitment is impossible. That doesn't undermine quantum cryptography, which can still be made secure for all practical purposes. But it exposes some of the limits of what is possible.

Well, so far so cryptic. What can be so fundamental about a desideratum for sending secure information? But as we saw earlier, the very existence of quantum information technologies such as quantum computing and quantum cryptography stems from deep features of quantum mechanics. It's not so much that these features lend themselves to applications in information processing (although that's true), but that quantum mechanics is increasingly – as Bub says – *about* information. What, then, we might express in terms of a protocol for sending secret data is actually a principle about what is knowable and what is not in the quantum world. The impossibility of unconditionally secure bit commitment in quantum mechanics in fact amounts to the condition that entangled states do not decay spontaneously as the particles move apart. That's to say, the correlation between some property of the two particles persists no matter how far apart they are.

Clifton, Bub and Halvorson showed that, if we make these three basic stipulations about quantum information, from them we can deduce a great many of the behaviours, such as superposition, entanglement, uncertainty and non-locality, at the heart of quantum theory. We don't get the *full* theory, but we get its essence. And these three principles in turn are related to the fact that quantum mechanics is a non-commutative kind of algebra.

All this is very much in the same spirit as the model of PR boxes described earlier, in which quantum-like (as well as super-quantum) non-local behaviour comes from rules about how information can be shared (or correlated) between different objects. Indeed, Bub suspects that the notion of information causality, which has been proposed as an axiom limiting the strength of that sharing to what we see in quantum mechanics (page 315), might also be at the root of the non-commutativity of quantum-like theories. Information causality, you might recall, imposes a constraint on the relationship between what can and cannot be deduced about a quantum system based on what has been already measured.

•

These ideas are still tentative and speculative. But they hint at an emerging story in which quantum mechanics has the features it does because there are certain things that cannot be done with information.

Why can't they be done? That would be like asking why the laws of physics take the form they do, and not some other. It's a valid question, but it can't be answered simply by appealing to those laws themselves. If what we're trying to do is to understand quantum mechanics, this amounts to figuring out what are the fundamental principles that give rise to its multitude of often counter-intuitive properties. Bub and colleagues are not saying that their three no-go rules about information *are* those fundamental principles – one can't, after all, deduce everything about quantum mechanics from these rules. But they are suggesting that a reconstructed quantum mechanics should look something like this: a set of rules for representing and manipulating information. You can, if you like, reformulate those rules in terms of waves and particles, wavefunctions and entanglement and 'measurement problems' – but you don't thereby add anything new in terms of what the theory can predict or explain about nature. You get a useful mathematical calculus for solving problems, and that's fine. But you'd best not try to attach much 'meaning' to it.

Still, we can speculate about where some of these constraints on the properties of information might come form. As a candidate for a simple unifying principle of quantum mechanics, Časlav Brukner and Anton Zeilinger have offered the notion that every fundamental constituent of a system* can only encode one bit of information: it

* This does not mean every 'fundamental particle', such as protons, quarks or electrons. They are clearly more complex than this, encoding more than one bit of information. The notion, rather, is that these known particles are composed of more primitive units of some (currently unknown) description.

can be this, or it can be that, and nothing else. After all, if it were any more complex than that, would it really be so fundamental? Quantum mechanics then emerges from a mismatch between the actual information-carrying capacity of the basic units of stuff and our beliefs about what they *ought* to be able to encode. If all the information-bearing potential is used up in answering particular questions, anything else we might try to measure is simply random. If the information content of a particle is all devoted to maintaining a correlation of some property with another particle, say, then the rest is random – and we are forced to make do with probabilities and with statements that are only statistically true.

The 'one bit per elementary particle' idea might or might not be true, but it offers a way to think about the issue on which quantum mechanics seems to pronounce very clearly: we can't know everything. The maximum information we can possibly know about a quantum system is not the complete information that specifies everything about how that system might behave. And the residue is not simply unknown; it is unspecified. We should probably not say that those other properties don't have fixed values, but rather that, being unmeasured, *they are not even properties*. And there is no absolute way to specify, before we look, which properties of a system have predictable, measureable values and which do not.*

* There's a philosophical question hanging over this discussion, which I will simply park here: if a question turns out to have no meaningful answer – if the answer is unspecified – are we then justified in still considering it a meaningful question in the first place?

It depends on how we ask the question – which is to say, on how we do the experiment. It is contextual.

•

If we are ever to be able to tell a 'simple story' about quantum mechanics, we can't avoid the Big Question: to what extent (if at all) the theory pronounces on the 'nature of reality'. Does quantum mechanics describe something that is really *out there* (is it an *ontic* theory – see page 54), or does it merely describe what we can know about the world (is it an *epistemic* theory)?

Ontic theories, such as hidden-variables models and the de Broglie–Bohm interpretation, take the view that quantum objects have objective properties – which means in turn that wavefunctions are 'real' entities, having a one-to-one correspondence with properties that don't depend on their being measured. The Copenhagen Interpretation, on the other hand, is epistemic, insisting that it's not physically meaningful to look for any layer of reality beneath what we can measure.

God plays dice only in an epistemic picture. In an ontic view, it only looks that way because we don't (and perhaps can't) know everything. Or (this option is not acknowledged enough) maybe quantum mechanics is a theory of yet another kind again, rooted perhaps not in some rather vague notion of 'information about the world' but in experience *of* the world. Certainly, philosophers have a thing or two to tell physicists about how delicate and slippery a term 'the nature of reality' is.

•

Some reconstructionists suspect that ultimately the correct description of the quantum world will prove to be ontic rather than epistemic: it will once more remove the human observer from the scene and return us to an objective view of reality. Others disagree, keeping faith with Bohr's insistence that quantum mechanics tells us not about the world but about what we can know of it.

How can we judge who's right, if the aim of all these models and theories is simply to return us to the same body of observed phenomena? It's not impossible that a reconstruction might actually predict new observable effects, and so could be subject to a specific experimental test. But Hardy thinks that the real criteria for success of a reconstruction must be theoretical. Do we get a better understanding of quantum theory? Do the axioms of a particular model give us new ideas about going beyond current physics, for example to develop the elusive quantum theory of gravity?

Even if no quantum reconstruction succeeds in finding a universally accepted set of principles that works, it won't be a wasted effort. Already these efforts have broadened our view. They suggest that our universe is just one of many mathematical possibilities about how information is distributed and made accessible, based in a description of events that is probabilistic rather than deterministic. And the challenge is then to find principles that single out quantum mechanics from the other options.

If we can find them, quantum mechanics will look a lot less mysterious, and we might hope that many of the outstanding questions will not so much acquire answers as simply evaporate.

The question is: what will we then be left with?

Can we ever get to

the bottom of it?

Well, perhaps now you can see the problem. The Spanish physicist Adán Cabello has put it rather nicely in this imagined scenario:

> Motivated by some recent news, a journalist asks a group of physicists: 'What's the meaning of the violation of Bell's inequality?' One physicist answers: 'It means that non-locality is an established fact.' Another says: 'There is no non-locality; the message is that measurement outcomes are irreducibly random.' A third one says: 'It cannot be answered simply on purely physical grounds, the answer requires an act of metaphysical judgement.' Puzzled by the answers, the journalist keeps asking questions about quantum theory: 'What is teleported in quantum teleportation?' 'How does a quantum computer really work?' Shockingly, for each of these questions, the journalist obtains a variety of answers which, in many cases, are mutually exclusive. At the end of the day, the journalist asks: 'How do you plan to make progress if, after ninety years of quantum theory, you still don't know what it means? How can you possibly identify the physical principles of quantum theory or expand quantum theory into gravity if you don't agree on what quantum theory is about?'

Having sometimes found myself in the shoes of that journalist, I'm inclined to agree. However, I don't find the situation frustrating, nor does it seem hopeless. On the contrary, the struggle is exciting and the progress promising – once we dispense with the clichés that journalists, often egged on by researchers, have wielded for too long.

It doesn't seem impossible that John Wheeler's dream – that we'll find deep laws of quantum mechanics which can be expressed in words anyone can grasp – will come to pass. But if we do not, that might not be because we simply fail to discover what those laws are. It could be far more interesting, and more unsettling, than that.

When it's said that quantum mechanics is 'weird', or that nobody understands it, the image tends to invite the analogy of a peculiar person whose behaviour and motives defy obvious explanation. But this is too glib. It's not so much understanding or even intuition that quantum mechanics defies, but our sense of logic itself. Sure, it's hard to intuit what it means for objects to travel along two paths at once, or to have their properties partly situated some place other than the object itself, and so on. But these are just attempts to express in everyday words a state of affairs that defeats the capabilities of language. Our language is designed to reflect the logic we're familiar with, but that logic won't work for quantum mechanics.

I do think that we can and will find better fundamental axioms for quantum mechanics, and I think they will be axioms about how information can exist and be discovered in the world. But it seems very likely that those axioms won't make 'sense' in a conventional way. To get the full picture, we need to accept what appear to be contradictory things. That is what Bohr was driving at with his notion of complementarity, although it was too vague and misdirected to express the whole truth. An affront to our sense of what should and should not be possible is probably never going to go away, not even if quantum mechanics is found to be only an approximation of some deeper theory.

You could put it this way: what is more fundamental, a fact established by logic or a fact established by observation? Everything that looks strange about quantum mechanics stems from the incompatibility of those two options.

The point is strikingly made in a thought experiment proposed by Yakir Aharonov, Sandu Popescu and their co-workers, which violates what they call the 'pigeonhole principle': if you put three (intact) pigeons into two pigeonholes, at least two of the pigeons must end up in the same hole. This principle, they say, captures 'the very essence of counting'. And yet for quantum particles – Aharonov and colleagues consider electrons, dispatched in a trio with parallel trajectories into a double-slit-style split-path apparatus – this need not be true. The scenario recalls the three boxes of Ernst Specker's Assyrian seer (page 191), and the reason for the outcome – to the extent that reason can be applied to such a paradoxical situation – is somewhat analogous: asking which box each of a given pair of particles is in delivers answers that need not be logically compatible with asking if the two particles are in the same box. There is no 'fact of the matter' outside the context in which that fact is interrogated.

•

This destabilization of facts is one of the most challenging aspects of quantum mechanics. How can we do science if the status of facts becomes uncertain and relative?

'Fact' was originally a legal term, etymologically derived from the Latin word for an action: it was a 'thing that was done', not some pre-existing truth. That might remain a useful distinction for quantum mechanics; certainly it seems to be what Niels Bohr had in mind in yoking facts to experiments. If I observe that something happened, and I can show that my observation is reliable, then surely it must be considered a fact? And if it's a fact, then by definition it must be true, right?*

But what, then, are the 'facts' of the double-slit experiment? That the particle travels along one route, or two? If we don't measure the particle's path, we seem compelled by observation of the outcome (namely, interference) to state as a fact that it travels both routes at once. This is the logical implication. If we measure the path, we find that it takes only one of them (and there is no longer interference). But because these are two different experiments, Bohr insisted that there was no problem here: we've no reason to expect the same answer. In this view, asking 'What are the facts of the double-slit experiment?' is an incomplete question.

Roland Omnès has sharpened Bohr's position by arguing that the concepts of 'fact' and 'truth' can only

* This is at root why the Many Worlds Interpretation of quantum mechanics is fundamentally different from any other theory in science, because it denies this statement. Its defenders might reply that what 'I' observe is only one of several 'facts' about the event. But since those other 'facts' may directly contradict the one I observe, none can meaningfully be said to be true. Neither can we even say that 'I' observed anything in the first place. The Many Worlds Interpretation denies language, but gets away with it because language has a notorious capacity to express things that appear to have meaning yet do not.

apply at the macroscopic scale, since only there can we truly observe anything. What we take to be common-sense logic and criteria for truth are in fact only criteria that (usually) emerge at the everyday scale – and which we can now understand and explain to an impressive degree as the consequence of quite different rules at other scales. Not just measurements and observations but even facts and familiar logical principles become possible *by degrees* – and are only fully well defined at the classical scale.

What makes these facts significant, Omnès says, is their *uniqueness*. If there is a fact about a situation, there can't be another fact that contradicts it.

Yet quantum mechanics can't tell us what these unique facts will be, only that they will be consistent with the *statistics* it predicts. If quantum theory says that an event has two possible outcomes with 50:50 probability, facts about its outcome in specific cases will accumulate in (roughly) this ratio. If quantum theory predicts that an outcome has zero probability, we'll never observe it as a fact.*

Are we really content to remain, as Bohr would have us do, at this level of observations, where facts are drawn from a statistical distribution created by unknowable mechanisms? Must we accept the Copenhagen prohibition on speaking of (although *not* necessarily denying) an underlying reality?

* A QBist (page 120) would disagree, saying instead only that we have then no reason to *expect* to observe it. To a QBist, quantum probabilities don't constrain the world to behave a certain way but speak only to an observer's expectations about that.

Let's take care: 'reality' is bandied about too recklessly. In everyday usage it's an inherently macroscopic concept: we can only view it through the lens of what we experience. In this sense, we have absolutely no reason to expect that it is 'reality all the way down'.

Still, almost all of science works fine by assuming that our perception of reality can be related to an underlying physical, tangible substrate *that doesn't innately depend on that perception*. We can account for the properties of stuff we touch, taste, smell and so forth by appealing to the concepts of atoms and molecules, and then more finely to protons and electrons and so on interacting by quantum rules. We have learnt to expect that we can explain experience through logical reasoning applied to what we can measure in ever more refined detail.

Quantum mechanics shows the limits of that approach: the places where our conventional, intuitive logic ultimately fails. It doesn't even have to be a microscopic limit, but just any place where quantum rules don't generate some classical approximation. In that regime, says Omnès, we can't any longer talk about a 'reality'. For him, reality must be a space in which facts are unique: in which, you might say, there are *events*. The rest is beyond our powers of reason. We simply can't bridge the gap, or at least not with quantum theory alone. As Berthold-Georg Englert puts it, the theory can't help us when we ask the question 'Why are there events?' All it can do is to show us that, to our surprise, this is a valid and puzzling question at all.

If that seems like an admission of defeat, says Omnès, it needn't. The very triumph of quantum mechanics is in having reached the point at which we must leave behind any notion of 'physical realism': the assumption that scientific investigation gives us access to and knowledge of physical reality. That assumption has in truth been more fraught in the history of science than is sometimes acknowledged: until Galileo forced the issue, Copernican theory sustained a delicate coexistence with Christian doctrine by being presented only as a manner of speaking, not a description of physical reality. But quantum mechanics shows that science itself ultimately disrupts the realist view: as Bohr put it, the theory requires 'a radical revision of our attitude toward the problem of physical reality'. This, says Omnès, is because quantum mechanics can't in itself connect us to conventional ideas of 'facts' – not without some extra assumptions. And he asks: isn't our arrival at this limit to knowledge of reality worth celebrating, not lamenting?

Perhaps so. But the mystery is that our equations can continue into this realm beyond realism and even thrive there, though we can't then deduce (or express) their meaning. It's not surprising, then, that some scientists want to make math itself the ultimate reality, a kind of numinous fabric from which all else emerges. Right now this may be a matter of taste. But when physicists – Everettians are particularly keen on this – exhort us to not get hung up on all-too-human words, we have a right to resist. Language is the only vehicle we have for constructing and conveying meaning: for talking about our universe. Relationships between numbers are no substitute. Science deserves more than that.

•

If intuitively transparent logic fails and math is too abstract a substitute, how else can we hope to get a handle on what quantum mechanics is 'telling us'? John Bell had something characteristically mischievous to say about that. 'Is it not good to know what follows from what even if it is not necessary for all practical purposes?' he asked. 'Suppose for example that quantum mechanics were found to resist precise formulation'–

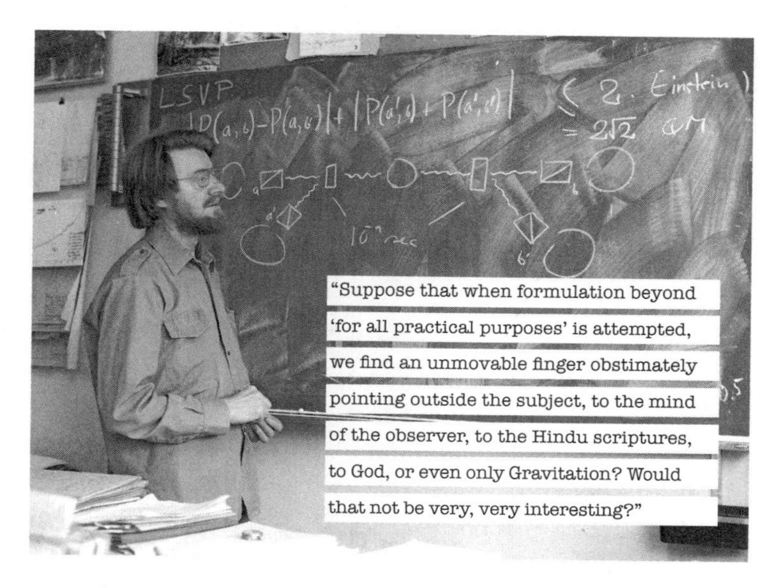

Indeed it would. But it seems unlikely that we will be forced to take recourse in holy scripture. Something more is needed, but that might turn out to be nothing else than better words. No one can yet say whether or not there are, beyond quantum theory's forbidding technical core, simple statements that express, crisply and clearly, what the machinery is about.

All we have right now are hints and guesses. And at this stage, to try to bring them into sharper focus is a risky business, which may demand an almost poetic level of expression that will (I fear) send some physicists scurrying for cover. Take, for example, Chris Fuchs' suggestion that

> The world is sensitive to our touch. It has a kind of 'Zing!' that makes it fly off in ways that were not imaginable classically. The whole structure of quantum mechanics may be nothing more than the optimal method of reasoning and processing information in the light of such a fundamental (wonderful) sensitivity.

What Fuchs means here is not that the human observer disturbs the world, in the way that Heisenberg misrepresented quantum uncertainty as a disturbance wrought by a high-magnification microscope. Rather, the microscopic world is sensitive to interactions of *any* kind: it is, you might say, exquisitely highly strung. And if it is wired that way, our interventions as active agents matter. Quantum mechanics is the machinery we humans need – at scales pitched midway between the subatomic and the galactic – to try to compile and quantify information about a world of that nature. It embodies what we have learnt about how to navigate in such a place.

Saying that our presence matters to what we see is a deeper idea than the common cliché that quantum mechanics makes the world dependent on the observer. For one thing, when we get hung up on the observer, we run into problems about what happens *when we don't observe*: we are back with Einstein's moon and Schrödinger's cat, or with superpositions that separate into different universes. Perhaps what we should be saying is not that quantum mechanics gives us a glimpse of what the universe is like and that this likeness depends on whether we intervene or not. Rather, quantum mechanics may be the theory we need *specifically to deal with interventions* in a particular kind of nature.

This is what the quantum interpretation called QBism (page 120) is really about, and it is why we would be wrong to regard it as some kind of solipsistic idea that 'it's all about us', or (worse) that 'reality is an illusion'. QBism is, rather, an expression of what Wheeler called the 'participatory universe', in which we play a role in the reality that we experience, without claiming that this is the whole story.

This is a fundamentally *realist* view at least to the extent that it allows stuff to happen without us. Pieces of the world come together, and facts arise out of that. We can't (yet?) say exactly how, let alone why, that happens. One might choose to identify this limitation with the 'intrinsic randomness' that has been ascribed to the quantum world, although Fuchs prefers to see it as a genuine autonomy: a 'creativity or novelty in the world'. Wheeler called it, rather cryptically (but not mystically) 'law without law'. In this view, laws enter into the universe only when (and because) we intervene. They are the probabilistic laws we have discovered to be effective in quantum mechanics, and which can become deterministic ones at scales where averages rule.

Wheeler developed a wonderful metaphor for illustrating this idea of a participatory realism: for showing how answers about 'reality' can emerge from the questions we ask in a way that is perfectly consistent, rule-bound and non-random without requiring a pre-existing 'truth'. You probably know the guessing game of Twenty Questions, in which one player leaves the room and the others agree on a word, or a person, or an object. Then the questioner returns and asks questions to which the only permitted answers are 'Yes' and 'No'. (You see: now you realize that this is a quantum game!)

Imagine you're the questioner. You start your questions, and receive the answers – but you find that after a few questions the answers take longer and longer to arrive. That's odd. Still, you sense you're closing in on the word, and finally you are sure you've got it: 'A cloud!' And everyone laughs and tells you you're right.

But then they explain what was going on. While you were out of the room, they didn't decide on that word at all. They didn't decide on *any* particular answer. The rule was simply that each person had to make sure that the answer they gave was consistent with the previous ones in fitting with *something*. And so the options became more and more constrained as the questions proceeded, and it took ever longer to figure out which word would still work. And everyone was forced, *by the nature of the questions*, to converge on the same word. If you had asked different questions, you'd have ended up with a different word: the answer was contextual. So there never was a preordained answer – you brought it into being, and in a way that is fully consistent with all the questions you asked. What's more, the very notion of an answer only makes sense when you play the game. It is meaningless to ask what the chosen word is without asking questions about it. Until you do that, there are only words.

•

How do you call this participatory world into being? With *information*! Or as Wheeler put it, you get 'It From Bit'.

Yes, yes. But information *about what*? What's the stuff, the fabric?

That might not fall within the jurisdiction of quantum mechanics at all. It doesn't mean that there *is* no stuff – I'm not sure what actual meaning attaches to the assertion sometimes made that 'the world *is* information'. Rather, we might need to say that it doesn't matter, for the interpretation of quantum mechanics (though it could matter a great deal for physics more broadly), what *stuff is*. We might as well regard it as particles. Why not? The issue is how the information about that stuff responds to our attempts to find out about it. Because in the end, that's all we can do in science: to try to find out.

I believe that most if not all of the thinking I have discussed in this book has converged, in one way or another, on this question: what is and is not permitted about information? We've seen how making a few assumptions about the answers – ones that are derived from how we have observed quantum systems to behave, and not from any preordained logic or intuition based on the classical world – can give rise to types of behaviour that at least resemble what we see in quantum mechanics.

I suspect it might be better still not to talk about quantum information – because 'information' might seem to imply something that is out there, if only we could get at it – but about quantum knowledge. Quantum mechanics is a theory of what is and is not knowable, and how those knowns are related. It is not yet able to say – and might never be able to say – where what is known *comes from*.

There's another important issue here too: what is the objective status of the information we *do* know (because we looked)? Is it available to all? Or is it subjective, specific to a certain observer looking in a certain way at a particular place or time in an attempt to find stuff out? Can I know what you know? Must what I know even be *compatible* with what you know? We just, well, don't know. But we can feel sure that this is going to be one of the big questions for future efforts to understand what quantum mechanics is telling us.

Perhaps you see the problem here – the problem with which all talk of 'quantum mechanics as information' must wrestle. We're used to the notion of things that in some sense *contain* information: books, computer memories, messages left on an answerphone. And we're used to the idea that we can *possess* information: I can know your email address, say. And these seem distinct: one is potential knowledge, the other actual knowledge, culled from potential knowledge according to our individual capacity. But quantum mechanics seems to make the interaction two-way: knowledge we possess affects what is knowable (and to others, or just to us?). Yes, it's confusing. But *that* is surely the right confusion to embrace, if we want to grapple with what this wonderful theory means.

I like to think of this informational perspective in terms of a distinction between a theory of Isness and a theory of Ifness. Quantum mechanics doesn't tell us how a thing *is*, but what (with calculable probability) it *could be*, along with – and this is crucial – a logic of the relationships between those 'coulds'. If *this*, then *that*.

What this means is that, to truly describe the features of quantum mechanics, as far as that is currently possible, we should replace all the conventional 'isms' with 'ifms'. For example:

Not
'here it is a particle, there it is a wave'
but
'if we measure things like this, the quantum object behaves in a manner we associate with particles; but if we measure it like that, it behaves as if it's a wave'

Not
'the particle is in two states at once'
but
'if we measure it, we will detect this state with probability X, and that state with probability Y'

This Ifness is perplexing, because it is not what we've come to associate with science. We're used to science telling us how things *are*, and if 'Ifs' arise, that's just because of our partial ignorance. But in quantum mechanics, Ifs are fundamental.

Is there an Isness beneath the Ifness? That's possible – and simply admitting as much takes us beyond the simplistic view of the Copenhagen Interpretation according to which there is nothing meaningful to be said beyond the results of observation. But even if there is, it will not be like the Isness of everyday life, in which objects have intrinsic, non-contextual, localized properties. It will not be a 'common sense' Isness.

Where we still struggle is in deciding what degree of reality, if any, to attribute to the Ifs of quantum mechanics. But perhaps we shouldn't get too hung up on that.

It's hard, after all, to even think about that question except in terms of Isness, and yet we have absolutely no reason (beyond our fallible intuition) to suppose that the universe must ultimately be an Is-world rather than an If-world. What's more, we don't need it to be so in order to recover the classical Is-world we experience. For we have a rather good understanding today of how classical Isness can and indeed must emerge from quantum Ifness. One of the most immediate and pressing questions now is to understand why the quantum Ifness has the particular character it does, and not some other. Maybe, if we can answer that question, we'll find a clue about how best to phrase the next one.

At any rate, it's vital that we understand this Ifness doesn't imply that the world – *our* world, our home – isn't holding anything back from us. It's just that classical physics has primed us to expect too much from it. We have become accustomed to asking questions and getting definite answers: What colour is it? How heavy? How fast? Forgetting the almost ludicrous amount about everyday objects of which we remain ignorant, we figured we could go on forever asking and being answered, at ever finer scales. When we discovered that we cannot, we felt short-changed by nature and pronounced it 'weird'.

That won't do any more. Nature does its best, and we need to adjust our expectations. It is time to go beyond weird.

Acknowledgements

I have benefitted tremendously from discussions with Cyril Branciard, Časlav Brukner, Andrew Cleland, Yves Colombe, Matthew Fisher, Chris Fuchs, Dagomir Kaszlikowski, Johannes Kofler, Anthony Laing, Franco Nori, Andrew Parry, Sandu Popescu, Ruediger Schack, Maximilian Schlosshauer, Luca Turin, Philip Walther and Wojciech Zurek, without whom the errors in this book would be substantially multiplied. They have reinforced my conviction that, collectively, scientists are the most generous academics in the world.

Editors who have a feeling for books as both physical and intellectual objects are of incomparable value to a writer, and I have been very fortunate to work with Jörg Hensgen and Stuart Williams, whose wise editorial advice has given the book the right shape and style. I would be lost without the guidance and support of my agent Clare Alexander. And I am always relieved to be in the reliable copy-editing hands of David Milner.

I should like to dedicate the book to two friends who did their best to teach me quantum mechanics many years ago. My gratitude to Peter Atkins comes largely from the tremendous support he has given to my writing on chemistry; but once upon a time I was in the audience of Oxford undergraduates to whom Peter delivered dazzling lectures on the topic. His style and clarity have, if anything, only improved over the years. Balázs Györffy's warmth and enthusiasm more than compensated for the fact that he struggled to bring the level of his delivery down to that of a baffled young chemistry graduate. He died in 2012, and his enormous generosity and energy are still sorely missed in the Bristol physics department.

Mei Lan and Amber showed me that children are open to anything, even quantum entanglement, and that they are our hope. And I thank them for loaning me their Sylvanian animals to assist me in my public talks on quantum mechanics.

Philip Ball
London, October 2017

Notes

To encounter the quantum is to feel: J. A. Wheeler, 'Delayed-choice experiments and the Bohr-Einstein dialogue', in *At Home in the Universe*, p. 130. American Institute of Physics, Woodbury NY, 1994.

Somewhere in [quantum theory] the distinction: E. T. Jaynes, 'Quantum Beats', in A. O. Barut (ed.), *Foundations of Radiation Theory and Quantum Electrodynamics*, p. 42. Plenum, New York, 1980.

We must never forget that 'reality' too: quoted in Wheeler & Zurek (1983), p. 5.

[Quantum mechanics] is a peculiar mixture describing in part: E. T. Jaynes, in W. H. Zurek (ed.), *Complexity, Entropy, and the Physics of Information*, p. 381. Addison-Wesley, New York, 1990.

Arguably the most important lesson of quantum mechanics: Aharonov et al. (2016), p. 532.

I hope you can accept nature as she is – absurd: R. Feynman, *QED: The Strange Theory of Light and Matter*, p. 10. Penguin, London, 1990.

. . . and this is a book about it: Historians of science will recognize this homage to Steven Shapin's provocative and influential book *The Scientific Revolution*, University of Chicago Press, Chicago, 1996.

I was born not understanding quantum mechanics: M. Feynman (ed.), *The Quotable Feynman*, p. 329. Princeton University Press, Princeton, 2015.

We can't pretend to understand it: ibid., p. 210.

I want to recapture that feeling for all: quoted in C. A. Fuchs, 'On participatory realism', in I. T. Durham & D. Rickles (eds), *Information and Interaction: Eddington, Wheeler, and the Limits of Knowledge*, p. 114. Springer, Cham., 2016.

We are suspended in language: quoted in A. Peterson, *Quantum Physics and the Philosophical Tradition*, p. 188. MIT Press, Cambridge MA, 1968.

If a man does not feel dizzy: quoted by C. A. von Weizsäcker, in M. Drieschner (ed.), *Carl Friedrich von Weizsäcker: Major Texts in Physics*, p. 77. Springer, Cham., 2014.

Sometimes the sensation persists for many minutes: Mermin (1998).

for anyone who ever regretted not taking physics: Susskind & Friedman (2014), dust jacket.

It looks completely crazy: Farmelo (ed.) (2002), p. 22.

the fuel of a machine that manufactures probabilities: Omnès (1999), p. 155.

The natural laws formulated mathematically: W. Heisenberg, *The Physicist's Conception of Nature*, transl. A. J. Pomerans, p. 15. Hutchinson, London, 1958.

a fundamental limitation of the program: A. Zeilinger, 'On the interpretation and philosophical foundation of quantum mechanics', in U. Ketvel et al. (eds), *Vastakohtien todellisuus*, Festschrift for K. V. Laurikainen, p. 5. Helsinki University Press, Helsinki, 1996.

There is no quantum world: quoted in A. Petersen, 'The Philosophy of Niels Bohr', *Bulletin of the Atomic Scientists* 19 (1963), 12.

an actor in [the] interplay between man: Heisenberg (1958), op. cit., p. 29.

mental concepts are the only reality: quoted in K. Ferguson, *Stephen Hawking: An Unfettered Mind* p. 433. St Martin's Griffin, London, 2017.

an experimental physics whose modes: Omnès (1994), p. 147.

In actuality it is wrong to talk of the 'route' of the photon: J. A. Wheeler, 'Law without law', in Wheeler & Zurek (eds) (1983), p. 192.

It is not at all the act of physical interaction: C. F. von Weizsäcker (1941). 'Zur Deutung der Quantenmechanik', *Zeitschrift für Physik* 118, 489–509. Translated in Ma *et al.* 2016.

No phenomenon is a phenomenon until: Wheeler (1983), op. cit.

Our task is to learn to use these words correctly: quoted by J. Kalckar, 'Niels Bohr and his youngest disciples', in S. Rozental (ed.), *Niels Bohr: His Life and Work as Seen by His Friends and Colleagues*, p. 234. North-Holland, Amsterdam, 1967.

Bohr's writings are characterized: C. F. von Weizsäcker, in Bastin (ed.) (1971), p. 33.

Bohr was essentially right but he did not know why: ibid., p. 28.

The Copenhagen Interpretation got to the top: Cushing (1994), p. 133.

wrong and misleading to break: D. Bohm, *Thought as a System*, p. 19. Routledge, London, 1994.

We could leave it at that: Englert (2013) [arxiv], p. 12.

There is no logical necessity: Fuchs & Peres (2000), p. 70.

the characteristic trait of quantum mechanics: Schrödinger (1935), p.555.

It has not come out as well: quoted in Harrigan & Spekkens (2007), p. 11.

I am therefore inclined to believe: quoted in Mermin (1985), p. 40.

I am a quantum engineer: quoted in Gisin (2001), p. 199.

Unperformed experiments have no results: A. Peres, 'Unperformed experiments have no results', *American Journal of Physics* 46 (1978), 745.

it presents us with a set of correlations: N. D. Mermin, *Boojums All the Way Through: Communicating Science in a Prosaic Age*, p. 174. Cambridge University Press, Cambridge, 1990.

having in it the living and the dead cat: E. Schrödinger, *Die Naturwissenschaften* 23 (1935), 807, 823, 844; English translation J. D. Trimmer, *Proceedings of the American Philosophical Society* 124 (1980), 323. Reprinted in Wheeler & Zurek (eds) (1983), p. 157.

It is typical of these cases that an indeterminacy: ibid.

if you want to make a simulation of nature: Feynman (1982), p. 486.

offers an unbreakable method for code-makers: Brassard (2015), p. 9.

This amounts to saying that whatever: W. H. Zurek, personal communication.

My own feeling on this issue: D. Gottesman, personal communication.

every quantum transition taking place: B. S. DeWitt, 'The many-universes interpretation of quantum mechanics', in B. d'Espagnat (ed), *Proceedings of the International School of Physics 'Enrico Fermi'*, Course IL: Foundations of Quantum Mechanics. Academic Press, New York, 1971.

all possible states exist at every instant: M. Tegmark, in *Scientific American*, May 2003. Reprinted in *Scientific American Cutting-Edge Science: Extreme Physics*, p. 114. Rosen, New York, 2008.

The only astonishing thing is that that's still controversial: Deutsch, in Saunders et al. (eds) (2010), p. 542.

gives an undue importance to the little differences: Omnès (1994), p. 347.

I feel a strong kinship with parallel Maxes: quoted in Hooper (2014).

because the measure of existence of worlds: Vaidman (2002).

The simple and obvious truth is that quantum phenomena: Peres (2002), pp. 1–2.

equations are ultimately more fundamental than words: Tegmark (1997), p. 4.

and it is like being in a holy city in great tumult: Fuchs (2002), p. 1.

literally a story, all in plain words: C. Fuchs, personal communication.

Despite all the posturing and grimacing: ibid.

fundamentally a theory about the representation: Bub (2004), p. 1.

however legitimate and necessary in application: Bell (1990), p. 34.

Motivated by some recent news: Cabello (2015), p. 1.

the very essence of counting: Aharonov et al. (2016), p. 532.

Is it not good to know what follows from what: Bell (2004), p. 214.

The world is sensitive to our touch: Fuchs (2002), p. 9.

Bibliography

'Arxiv' refers to the physics preprint server arxiv.org. These papers are accessible at http://arxiv.org/abs/[number].

Aczel, A. D. 2003. *Entanglement: The Greatest Mystery in Physics*. John Wiley, New York.

Aharonov, Y., Colombo, F., Popescu, S., Sabadini, I., Struppa, D. C. & Tollaksen, J. 2016. 'Quantum violation of the pigeonhole principle and the nature of quantum correlations', *Proceedings of the National Academy of Sciences USA* 113, 532–5.

Albert, D. 2014. 'Physics and narrative', in Struppa, D. C. & Tollaksen, J. (eds), *Quantum Theory: A Two-Time Success Story*, 171–82. Springer, Milan.

Al-Khalili, J. 2003. *Quantum: A Guide For the Perplexed*. Weidenfeld & Nicolson, London.

Arndt, M., Nairz, O., Vos-Andreae, J., Keller, C., van der Zouw, G. & Zeilinger, A. 1999. 'Wave-particle duality of C_{60} molecules', *Nature* 401, 680–2.

Aspect, A., Dalibard, J. & Roger, G. 1982. 'Experimental test of Bell's inequalities using time-varying analyzers', *Physical Review Letters* 69, 1804.

Aspect, A. 2015. 'Closing the door on Einstein and Bohr's debate', *Physics* 8, 123.

Aspelmeyer, M. & Zeilinger, A. 2008. 'A quantum renaissance', *Physics World*, July, 22–8.

Aspelmeyer, M., Meystre, P. & Schwab, K. 2012. 'Quantum optomechanics', *Physics Today*, July, 29–35.

Ball, P. 2008. 'Quantum all the way', *Nature* 453, 22–5.

Ball, P. 2013. 'Quantum quest', *Nature* 501, 154–6.

Ball, P. 2014. 'Questioning quantum speed', *Physics World* January, 38–41.

Ball, P. 2017. 'A world without cause and effect', *Nature* 546, 590–592.

Ball, P. 2017. 'Quantum theory rebuilt from simple physical principles', *Quanta* 30 August, www.quantamagazine.org/quantum-theory-rebuilt-from-simple-physical-principles 20170830/

Bastin, T. (ed.). 1971. *Quantum Theory and Beyond*. Cambridge University Press, London.

Bell, J. S. 1964. 'On the Einstein-Podolsky-Rosen paradox', *Physics* 1, 195–200.

Bell, J. S. 1990. 'Against measurement', *Physics World* August, 33–40.

Bell, J. S. 2004. *Speakable and Unspeakable in Quantum Mechanics: Collected Papers on Quantum Philosophy*. Cambridge University Press, Cambridge.

Bohm, D. & Hiley, B. 1993. *The Undivided Universe*. Routledge, London.

Bouchard, F., Harris, J., Mand, H., Bent, N., Santamato, E., Boyd, R. W. & Karimi, E. 2015. 'Observation of quantum recoherence of photons by spatial propagation', *Scientific Reports* 5:15330.

Branciard, C. 2013. 'Error-tradeoff and error-disturbance relations for incompatible quantum measurements', *Proceedings of the National Academy of Science USA* 110, 6742.

Brassard, G. 2005. 'Is information the key?', *Nature Physics* 1, 2.

Brassard, G. 2015. 'Cryptography in a quantum world', Arxiv: 1510.04256.

Brukner, Č. 2014. 'Quantum causality', *Nature Physics* 10, 259–63.

Brukner, Č. 2015. 'On the quantum measurement problem', Arxiv: 1507.05255.

Brukner, Č. & Zeilinger, A. 2002. 'Information and fundamental elements of the structure of quantum theory', Arxiv: quant-ph/0212084.

Bub, J. 1974. *The Interpretation of Quantum Mechanics*. Reidel, Dordrecht.

Bub, J. 1997. *Interpreting the Quantum World*. Cambridge University Press, Cambridge.

Bub, J. 2004. 'Quantum mechanics is about quantum information', Arxiv: quant-ph/0408020.

Buscemi, F., Hall, M. J. W., Ozawa, M. & Wilde, M. W. 2014. 'Noise and disturbance in quantum measurements: an information-theoretic approach', *Physical Review Letters* 112, 050401.

Busch, P., Lahti, P. & Werner, R. F. 2013. 'Proof of Heisenberg's error-disturbance relation', *Physical Review Letters* 111, 160405.

Cabello, A. 2015. 'Interpretations of quantum theory: a map of madness', Arxiv: 1509.04711.

Castelvecchi, D. 2015. 'Quantum technology probes ultimate limits of vision', *Nature News*, 15 June. http://www.nature.com/news/quantum-technology-probes-ultimate-limits-of-vision-1.17731.

Chiribella, G. 2012. 'Perfect discrimination of no-signalling channels via quantum superposition of causal structures', *Physical Review Letters A* 86, 040301(R).

Clauser, J. F., Horne, M. A., Shimony A. & Holt, R. A. 1969. 'Proposed experiment to test local hidden-variable theories', *Physical Review Letters* 23, 880–4.

Clifton, R., Bub, J. & Halvorson, H. 2003. 'Characterizing quantum theory in terms of information-theoretic constraints', *Foundations of Physics* 33, 1561.

Cox, B. & Forshaw, J. 2011. *The Quantum Universe: Everything That Can Happen Does Happen*. Allen Lane, London.

Crease, R. & Goldhaber, A. S. 2014. *The Quantum Moment*. W. W. Norton, New York.

Cushing, J. T. 1994. *Quantum Mechanics: Historical Contingency and the Copenhagen Hegemony*. University of Chicago Press, Chicago.

Deutsch, D. 1985. 'Quantum theory, the Church-Turing principle and the universal quantum computer', *Proceedings of the Royal Society A* 400, 97–117.

Deutsch, D. 1997. *The Fabric of Reality*. Penguin, London.

Devitt, S. J., Nemoto, K. & Munro, W. J. 2009. 'Quantum error correction for beginners', Arxiv: 0905.2794.

Einstein, A., Podolsky, B. & Rosen, N. 1935. 'Can quantum-mechanical description of physical reality be considered complete?', *Physical Review* 47, 777.

Englert, B.-G. 2013. 'On quantum theory', *European Physical Journal D* 67, 238. See Arxiv: 1308.5290.

Erhart, J., Sponar, S., Sulyok, G., Badurek, G., Ozawa, M. & Hasegawa, Y. 2012. *Nature Physics* 8, 185.

Everett, H. III. 1957. '"Relative state" formulation of quantum mechanics', *Reviews of Modern Physics* 29, 454.

Everett, H. III. 1956. 'The theory of the universal wave function', PhD thesis (long version). Available at http://ucispace.lib.uci.edu/handle/10575/1302.

Farmelo, G. (ed.). 2002. *It Must Be Beautiful: Great Equations of Modern Science*. Granta, London.

Falk, D. 2016. 'New support for alternative quantum view', *Quanta*, 16 May. https://www.quantamagazine.org/20160517-pilot-wave-theory-gains-experimental-support/.

Feschbach, H., Matsui, T. & Oleson, A. (eds). 1988. *Niels Bohr: Physics and the World*. Harwood Academic, Chur.

Feynman, R. 1982. 'Simulating physics with computers', *International Journal of Theoretical Physics* 21, 467–88.

Fuchs, C. A. & Peres, A. 2000. 'Quantum theory needs no "interpretation"', *Physics Today*, March, 70–1.

Fuchs, C. A. 2001. 'Quantum foundations in the light of quantum information', Arxiv: quant-ph/0106166.

Fuchs, C. A. 2002. 'Quantum mechanics as quantum information (and only a little more)', Arxiv: quant-ph/0205039.

Fuchs, C. A. 2010. 'QBism, the perimeter of Quantum Bayesianism', Arxiv: 1003.5209.

Fuchs, C. A. 2012. 'Interview with a Quantum Bayesian', Arxiv: 1207.2141.

Fuchs, C. A., Mermin, N. D. & Schack, R. 2014. 'An introduction to QBism with an application to the locality of quantum mechanics', *American Journal of Physics* 82, 749–54. See Arxiv: 1311.5253.

Fuchs, C. A. 2016. 'On participatory realism', Arxiv: 1601.04360.

Gerlich, S., Eibenberger, S., Tomandl, M., Nimmrichter, S., Hornberger, K., Fagan, P. J., Tüxen, J., Mayor, M. & Arndt, M. 2011. 'Quantum interference of large organic molecules', *Nature Communications* 2, 263.

Ghirardi, G. C., Rimini, A. & Weber, T. 1986. 'An explicit model for a unified description of microscopic and macroscopic systems', *Physical Review D* 34, 470.

Gibney, E. 2014. 'Quantum computer quest', *Nature* 516, 24.

Gisin, N. 2002. 'Sundays in a quantum engineer's life', in R. A. Bertlmann & A. Zeilinger (eds), *Quantum [Un]speakable*, 199–208. Springer, Berlin. See Arxiv: quant-ph/0104140.

Giustina, M. et al. 2013. 'Bell violations using entangled photons without the fair-sampling assumption', *Nature* 497, 227–30.

Greene, B. 2012. *The Hidden Reality: Parallel Universes and the Deep Laws of the Cosmos*. Penguin, London.

Gribbin, J. 1985. *In Search of Schrödinger's Cat*. Black Swan, London.

Grinbaum, A. 2007. 'Reconstruction of quantum theory', *British Journal for the Philosophy of Science* 58, 387–408.

Gröblacher, S. et al. 2007. 'An experimental test of non-local realism', *Nature* 446, 871–5.

Guérin, P. A., Feix, A., Araújo, M. & Brukner, Č. 2016. 'Exponential communication complexity advantage from quantum superposition of the direction of communication', *Physical Review Letters* 117, 100502.

Hackermüller, L., Hornberger, K., Brezger, B., Zeilinger, A. & Arndy, M. 2004. 'Decoherence of matter waves by thermal emission of radiation', *Nature* 427, 711–14.

Hardy, L. 2001. 'Quantum theory from five reasonable axioms', Arxiv: quant-ph/0101012.

Hardy, L. 2001. 'Why quantum theory?', Arxiv: quant-ph/0111068.

Hardy, L. 2007. 'Quantum gravity computers: on the theory of computation with indefinite causal structure', Arxiv: quant-ph/0701019.

Hardy, L. 2011. 'Reformulating and reconstructing quantum theory', Arxiv: 1104.2066.

Hardy, L. & Spekkens, R. 2010. 'Why physics needs quantum foundations', Arxiv: 1003.5008.

Harrigan, N. & Spekkens, R. W. 2007. 'Einstein, incompleteness, and the epistemic view of quantum states', Arxiv: 0706.2661.

Hartle, J. B. 1997. 'Quantum cosmology: problems for the 21st century', Arxiv: gr-qc/9210006.

Heisenberg, W. 1927. 'Über den anschaulichen Inhalt der quantentheoretischen Kinematik und Mechanik', *Zeitschrift für Physik* 43, 172–98.

Hensen, B. et al. 2015. 'Experimental loophole-free violation of a

Bell inequality using entangled electron spins separated by 1.3 km', Arxiv: 1508.05949.

Hooper, R. 2014. 'Multiverse me: Should I care about my other selves?', *New Scientist*, 24 September.

Hornberger, K., Uttenthaler, S., Brezger, B., Hackermüller, L., Arndt, M. & Zeilinger, A. 2003. 'Collisional decoherence observed in matter wave interferometry', Arxiv: quant-ph/0303093.

Howard, D. 2004. 'Who invented the "Copenhagen Interpretation"? A study in mythology', *Philosophy of Science* 71, 669–82.

Howard, M., Wallman, J., Veitch, V. & Emerson, J. 2014. 'Contextuality supplies the magic for quantum computation', *Nature* 510, 351–5.

Jeong, H., Paternostro, M. & Ralph, T. C. 2009. 'Failure of local realism revealed by extremely-coarse-grained measurements', *Physical Review Letters* 102, 060403.

Jeong, H., Lim, Y. & Kim, M. S. 2014. 'Coarsening measurement references and the quantum-to-classical transition. *Physical Review Letters* 112, 010402.

Joos, E., Zeh, H. D., Kiefer, C., Giulini, D. J. W., Kupsch, J. & Stamatescu, I.-O. 2003. *Decoherence and the Appearance of a Classical World in Quantum Theory*, 2nd edn. Springer, Berlin.

Kaltenbaek, R., Hechenblaikner, G., Kiesel, N., Romero-Isart, O., Schwab, K. C., Johann, U. & Aspelmeyer, M. 2012. 'Macroscopic quantum resonators', Arxiv: 1201.4756.

Kandea, F., Baek, S.-Y., Ozawa, M. & Edamatsu, K. 2014. *Physical Review Letters* 112, 020402.

Kastner, R. E., Jeknić-Dugić, J. & Jaroszkiewicz, G. 2016. *Quantum Structural Studies: Classical Emergence From the Quantum Level*. World Scientific, Singapore.

Kent, A. 1990. 'Against Many Worlds interpretation', *International Journal of Modern Physics A* 5, 1745.

Kent, A. 2014. 'Our quantum problem', *Aeon*, 28 January. https://aeon.co/essays/what-really-happens-in-schrodinger-s-box.

Kent, A. 2014. 'Does it make sense to speak of self-locating uncertainty in the universal wave function? Remarks on Sebens and Carroll', Arxiv: 1408.1944.

Kent, A. 2016. 'Quanta and qualia', Arxiv: 1608.04804.

Kirkpatrick, K. A. 2003. ' "Quantal" behavior in classical probability', *Foundations of Physics* 16, 199–224.

Knee, G. C. et al. 2012. 'Violation of a Leggett-Garg inequality with ideal non-invasive measurements', *Nature Communications* 3, 606.

Kofler, J. & Brukner, Č. 2007. 'Classical world arising out of quantum physics under the restriction of coarse-grained measurements', *Physical Review Letters* 99, 180403.

Kumar, M. 2008. *Quantum: Einstein, Bohr and the Great Debate About the Nature of Reality.* Icon, London.

Kunjwal, R. & Spekkens, R. W. 2015. 'From the Kochen-Specker theorem to noncontextuality inequalities without assuming determinism', Arxiv: 1506.04150.

Kurzynski, P., Cabello, A. & Kaszlikowski, D. 2014. 'Fundamental monogamy relation between contextuality and nonlocality', *Physical Review Letters* 112, 100401.

Lee, C. W. & Jeong, H. 2011. 'Quantification of macroscopic quantum superposition's within phase space', *Physical Review Letters* 106, 220401.

Leggett, A. J. & Garg, A. 1985. 'Quantum mechanics versus macroscopic realism: is the flux there when nobody looks?', *Physical Review Letters* 54, 857.

Li, T. & Yin, Z.-Q. 2015. 'Quantum superposition, entanglement, and state teleportation of a microorganism on an electromechanical oscillator', Arxiv: 1509.03763.

Lindley, D. 1997. *Where Does the Weirdness Go?* Vintage, London.

Lindley, D. 2008. *Uncertainty: Einstein, Heisenberg, Bohr, and the Struggle for the Soul of Science.* Doubleday, New York.

Ma, X.-S., Kofler, J. & Zeilinger, A. 2015. *'Delayed-choice gedanken experiments and their realization'*, Arxiv: 1407.2930.

Mayers, D. 1997. 'Unconditionally secure quantum bit commitment is impossible', Arxiv: quant-ph/9605044.

Merali, Z. 2011. 'Quantum effects brought to light', *Nature News*, 28 April. http://www.nature.com/news/2011/110428/full/news.2011.252.html.

Merali, Z. 2011. 'The power of discord', *Nature* 474, 24–6.

Mermin, N. D. 1985. 'Is the moon there when nobody looks? Reality and the quantum theory', *Physics Today*, April, 38–47.

Mermin, N. D. 1993. 'Hidden variables and the two theorems of John Bell', *Reviews of Modern Physics* 65, 803.

Mermin, N. D. 1998. 'The Ithaca interpretation of quantum mechanics', *Pramana* 51, 549–65. See Arxiv: quant-ph/9609013.

Musser, G. 2015. *Spooky Action at a Distance.* Farrar, Straus & Giroux, New York.

Nairz, O., Arndt, M. & Zeilinger, A. 2003. 'Quantum interference with large molecules', *American Journal of Physics* 71, 319–25.

Ollivier, H., Poulin, D. & Zurek, W. H. 2009. 'Environment as a witness: selective proliferation of information and emergence of objectivity in a quantum universe', *Physical Review A* 72, 423113.

Omnès, R. 1994. *The Interpretation of Quantum Mechanics.* Princeton University Press, Princeton.

Omnès, R. 1999. *Quantum Philosophy.* Princeton University Press, Princeton.

Oreshkov, O., Costa, F. & Brukner, Č. 2012. 'Quantum correlations with no causal order', *Nature Communications* 3, 1092.

Ozawa, M. 2003. 'Universally valid reformulation of the Heisenberg uncertainty principle on noise and disturbance in measurement', *Physical Review A* 67, 042105.

Palacios-Laloy, A., Mallet, F., Nguyen, F., Bertet, P., Vion, D., Esteve, D. & Korotkov, A. N. 2010. 'Experimental violation of a Bell's inequality in time with weak measurement', *Nature Physics* 6, 442–7.

Pawlowski, M., Paterek, T., Kaszlikowski, D., Scarani, V., Winter, A. & Zukowski, M. 2009. 'Information causality as a physical principle', *Nature* 461, 1101–4.

Peat, F. D. 1990. *Einstein's Moon: Bell's Theorem and the Curious Quest for Quantum Reality.* Contemporary Books, Chicago.

Peres, A. 1997. 'Interpreting the quantum world' (book review), Arxiv: quant-ph/9711003.

Peres, A. 2002. 'What's wrong with these observables?', Arxiv: quant-ph/0207020.

Pomarico, E., Sanguinetti, B., Sekatski, P. & Gisin, N. 2011. 'Experimental amplification of an entangled photon: what if the detection loophole is ignored?', Arxiv: 1104.2212.

Popescu, S. & Rohrlich, D. 1997. 'Causality and nonlocality as axioms for quantum mechanics', Arxiv: quant-ph/9709026.

Popescu, S. 2014. 'Nonlocality beyond quantum mechanics', *Nature Physics* 10, 264.

Procopio, L. M. et al. 2015. 'Experimental superposition of orders of quantum gates', *Nature Communications* 6, 7913.

Pusey, M. F., Barrett, J. & Rudolph, T. 2012. 'On the reality of the quantum state', Arxiv: quant-ph/1111.3328.

Riedel, C. J. & Zurek, W. H. 2010. 'Quantum Darwinism in an everyday environment: huge redundancy in scattered photons', *Physical Review Letters* 105, 020404.

Riedel, C. J., Zurek, W. H. & Zwolak, M. 2013. 'The objective past of a quantum universe – Part I: Redundant records of consistent histories', Arxiv: 1312.0331.

Ringbauer, M., Duffus, B., Branciard, C., Cavalcanti, E. G., White, A. G. & Fedrizzi, A. 2015. 'Measurements on the reality of the wavefunction', *Nature Physics* 11, 249–54.

Rohrlich, D. 2014. 'PR-box correlations have no classical limit', Arxiv:1407.8530.

Romero-Isart, O., Juan, M. L., Quidant, R. & Cirac, J. I. 2009. 'Towards quantum superposition of living organisms', Arxiv: 0909.1469.

Rowe, M. A., Kielpinski, D., Meyer, V., Sackett, C. A., Itano, W. M., Monroe, C. & Wineland, D. J. 2001. 'Experimental violation of a Bell's inequality with efficient detection', *Nature* 409, 791–4.

Rozema, L. A., Darabi, A., Mahler, D. H., Hayat, A., Soudagar, Y. & Steinberg, A. M. 2012. 'Violation of Heisenberg's measurement-disturbance relationship by weak measurements', *Physical Review Letters* 109, 100404.

Saunders, S., Barrett, J., Kent, A. & Wallace, D. (eds). 2010. *Many Worlds? Everett, Quantum Theory, and Reality*. Oxford University Press, Oxford.

Schack, R. 2002. 'Quantum theory from four of Hardy's axioms', Arxiv: quant-ph/0210017.

Scheidl, T. et al. 2010. 'Violation of local realism with freedom of choice', *Proceedings of the National Academy of Sciences USA* 107, 19708–13.

Schlosshauer, M. 2007. *Decoherence and the Quantum-to-Classical Transition*. Springer, Berlin.

Schlosshauer, M. (ed.). 2011. *Elegance and Enigma: The Quantum Interviews*. Springer, Berlin.

Schlosshauer, M. 2014. 'The quantum-to-classical transition and decoherence', Arxiv: 1404.2635.

Schlosshauer, M., Kofler, J. & Zeilinger, A. 2013. 'A snapshot of foundational attitudes toward quantum mechanics', *Studies in the History and Philosophy of Modern Physics* 44, 222–30.

Schlosshauer, M. & Fine, A. 2014. 'No-go theorem for the composition of quantum systems', *Physical Review Letters* 112, 070407.

Schreiber, Z. 1994. 'The nine lives of Schrödinger's cat', Arxiv: quant-ph/9501014.

Schrödinger, E. 1935. 'Discussion of probability relations between separated systems', *Mathematical Proceedings of the Cambridge Philosophical Society* 31, 555–63.

Schrödinger, E. 1936. 'Probability relations between separated systems', *Mathematical Proceedings of the Cambridge Philosophical Society* 32, 446–52.

Sorkin, R. D. 1994. 'Quantum mechanics as quantum measure theory', Arxiv: gr-qc/9401003.

Spekkens, R. W. 2007. 'In defense of the epistemic view of quantum states: a toy theory', *Physical Review A* 75, 032110.

Susskind, L. & Friedman, A. 2014. *Quantum Mechanics: The Theoretical Minimum*. Allen Lane, London.

Tegmark, M. 1997. 'The interpretation of quantum mechanics: many worlds or many words?', Arxiv: quant-ph/9709032.

Tegmark, M. 2000. 'Importance of quantum decoherence in brain processes', *Physical Review E* 61, 4194.

Tegmark, M. & Wheeler, J. A. 2001. '100 years of the quantum', Arxiv: quant-ph/0101077.

Timpson, C. G. 2004. Quantum information theory and the foundations of quantum mechanics. PhD thesis, University of Oxford. See arxiv: quant-ph/0412063.

Tonomura, A., Endo, J., Matsuda, T., Kawasaki, T. & Ezawa, H. 1989. 'Demonstration of single-electron buildup of an interference pattern', *American Journal of Physics* 57, 117–120.

Vaidman, L. 1996. 'On schizophrenic experiences of the neutron, or Why we should believe in the Many Worlds interpretation of quantum theory', Arxiv: quant-ph/9609006.

Vaidman, L. 2002, revised 2014. 'Many-Worlds Interpretation of quantum mechanics', in E. N. Zalta (ed.), *Stanford*

Encyclopedia of Philosophy. https://plato.stanford.edu/entries/qm-manyworlds/.

Vedral, V. 2010. *Decoding Reality: The Universe as Quantum Information*. Oxford University Press, Oxford.

Wallace, D. 2002. 'Worlds in the Everett interpretation', *Studies in the History and Philosophy of Modern Physics* 33, 637.

Wallace, D. 2012. *The Emergent Multiverse*. Oxford University Press, Oxford.

Weihs, G., Jennewein, T., Simon, C., Weinfurter, H. & Zeilinger, A. 1998. 'Violation of Bell's inequality under strict Einstein locality conditions', *Physical Review Letters* 81, 5039.

Wheeler, J. & Zurek, W. H. (eds). 1983. *Quantum Theory and Measurement*. Princeton University Press, Princeton.

Wootters, W. K. & Zurek, W. H. 2009. 'The no-cloning theorem', *Physics Today*, February, 76–7.

Zeilinger, A. 1999. 'A foundational principle for quantum mechanics', *Foundations of Physics* 29, 631–43.

Zeilinger, A. 2006. 'Essential quantum entanglement', in G. Fraser (ed.), *The New Physics*. Cambridge University Press, Cambridge.

Zeilinger, A. 2010. *Dance of the Photons*. Farrar, Straus & Giroux, New York.

Zukowski, M. & Brukner, Č. 2015. 'Quantum non-locality: It ain't necessarily so . . .', Arxiv: 1501.04618.

Zurek, W. H. (ed.). 1990. *Complexity, Entropy and the Physics of Information*. Addison-Wesley, Redwood City CA.

Zurek, W. H. 2003. 'Decoherence, einselection, and the quantum origins of the classical', Arxiv: quant-ph/0105127.

Zurek, W. H. 2005. 'Probabilities from entanglement, Born's rule $p_k=|\Psi_k|^2$ from envariance', Arxiv: quant-ph/0405161.

Zurek, W. H. 2009. 'Quantum Darwinism', Arxiv: 0903.5082.

Zurek, W. H. 2014. 'Quantum Darwinism, decoherence and the randomness of quantum jumps', *Physics Today*, October, 44.

Zurek, W. H. 1998. 'Decoherence, chaos, quantum-classical correspondence, and the algorithmic arrow of time', Arxiv: quant-ph/9802054.

Index